CW00727357

Building a Sustainable Political Economy: SPERI Research & Policy

Series editors

Colin Hay
SPERI
University of Sheffield
Sheffield, UK

Anthony Payne
SPERI
University of Sheffield
Sheffield, UK

The Sheffield Political Economy Research Institute (SPERI) is an innovation in higher education research and outreach. It brings together leading international researchers in the social sciences, policy makers, journalists and opinion formers to reassess and develop proposals in response to the political and economic issues posed by the current combination of financial crisis, shifting economic power and environmental threat. Building a Sustainable Political Economy: SPERI Research & Policy will serve as a key outlet for SPERI's published work. Each title will summarise and disseminate to an academic and postgraduate student audience, as well as directly to policy-makers and journalists, key policy-oriented research findings designed to further the development of a more sustainable future for the national, regional and world economy following the global financial crisis. It takes a holistic and interdisciplinary view of political economy in which the local, national, regional and global interact at all times and in complex ways. The SPERI research agenda, and hence the focus of the series, seeks to explore the core economic and political questions that require us to develop a new sustainable model of political economy. The SPERI research agenda, and hence the focus of the series, seeks to explore the core economic and political questions that require us to develop a new sustainable model of political economy at all times and in complex ways.

More information about this series at
http://www.springer.com/series/14879

Colin Hay · Tom Hunt
Editors

The Coming Crisis

Editors
Colin Hay
SPERI
University of Sheffield
Sheffield, UK

Tom Hunt
SPERI
University of Sheffield
Sheffield, UK

Building a Sustainable Political Economy: SPERI Research & Policy
ISBN 978-3-319-63813-3 ISBN 978-3-319-63814-0 (eBook)
DOI 10.1007/978-3-319-63814-0

Library of Congress Control Number: 2017948268

Cover illustration: Pattern adapted from an Indian cotton print produced in the 19th century

Printed on acid-free paper

This Palgrave Macmillan imprint is published by Springer Nature
The registered company is Springer International Publishing AG
The registered company address is: Gewerbestrasse 11, 6330 Cham, Switzerland

Contents

Editors and Contributors

About the Editors

Colin Hay is the Founding Co-Director of the Sheffield Political Economy Research Institute (SPERI) at the University of Sheffield and Professor of Political Science at Sciences Po, Paris. He is editor in chief of *New Political Economy* and founding co-editor of *Comparative European Politics and British Politics*. He is the author of numerous books including *Civic Capitalism* (Polity, 2015, with Anthony Payne), *The Legacy of Thatcherism* (Oxford University Press, 2014, with Stephen Farrall), *The Failure of Anglo-Liberal Capitalism* (Palgrave, 2013) and *The Political Economy of European Welfare Capitalism* (Palgrave, 2012, with Daniel Wincott).

Tom Hunt is a Policy Research Officer at the Sheffield Political Economy Research Institute (SPERI) at the University of Sheffield. He leads SPERI's partnership with the All-Party Parliamentary Group on Inclusive Growth and is the author of several SPERI British Political Economy research briefs.

Contributors

Andrew Baker is a Faculty Professorial Fellow in Political Economy at the University of Sheffield. He has published over 30 academic book chapters and articles on financial governance and has authored and co-authored two books, including *The Group of Seven*, (Routledge, 2006). From 2010 to 2015 he served as the senior managing editor of The British Journal of Politics and International Relations.

Jacqueline Best is Professor in the School of Political Studies at the University of Ottawa. She is the author of a number of books including *Governing Failure: Provisional Expertise* and the *Transformation of Global Development Finance* and *The Return of the Public in Global Governance*, both published by Cambridge University Press in 2014. She is an editor of the Review of International Political Economy.

Matthew Bishop is Senior Lecturer in International Politics at the University of Sheffield. He has published in *Review of International Studies, Review of International Political Economy* and the *Journal of Development Studies.* He is also the author of two books: *The Political Economy of Caribbean Development*, and, *Democratization: A Critical Introduction* (with Jean Grugel), both published by Palgrave Macmillan in 2013.

Martin Craig is a Research Fellow at the Sheffield Political Economy Research Institute at the University of Sheffield. His research addresses the political economy of Britain's 'environmental state'. He has published in *New Political Economy* and is the author of *Ecological Political Economy and the Socio-Ecological Crisis* (Palgrave, 2017).

Peter Dauvergne is Professor of International Relations at the University of British Columbia. His recent books include *Paths to a Green World*, (co-authored with Jennifer Clapp, MIT Press, 2011), *Eco-Business* (co-authored with Jane Lister, MIT Press, 2013), *Protest Inc.* (co-authored with Genevieve LeBaron, Polity Press, 2014), and *Environmentalism of the Rich* (MIT Press, 2016). He is the founding and past editor of the journal *Global Environmental Politics.*

Andrew Gamble is Professor of Politics at the University of Sheffield, Emeritus Professor of Politics at the University of Cambridge, and a Fellow of the British Academy and the Academy of Social Sciences. He is the author of many books on politics and political economy, including such *The Free Economy and the Strong State, Politics and Fate*, and *Crisis without End? The Unravelling of Western Prosperity.* His latest book—*Can the Welfare State Survive?*—was published in 2016. In 2005, he received the Isaiah Berlin Prize from the Political Studies Association for lifetime contribution to political studies.

Jeremy Green is a Lecturer in the Department of Politics and International Studies at the University of Cambridge. His work has been published in a number of academic journals including *New Political Economy, Review of International Studies* and *European Journal of International Relations.* He co-edited *The British Growth Crisis: The Search for a New Model* (Palgrave, 2015) with Colin Hay and Peter Gooby-Taylor.

Scott Lavery is a Research Fellow at the Sheffield Political Economy Research Institute (SPERI) at the University of Sheffield. His research examines the evolving relation between British and European capitalism. He has published in *New Political Economy, Transactions of the Institute of British Geographers* and the *British Journal of Politics and International Relations.* He is also the author of numerous SPERI papers and policy briefings.

Genevieve LeBaron is Senior Lecturer in Politics at the University of Sheffield and Human Trafficking and Modern Day Slavery Fellow at Yale University. She is the author of *Protest Inc.: The Corporatization of Activism* (Polity, 2014, with Peter Dauvergne) and 15 journal articles including in *Review of International Studies, Brown Journal of World Affairs, Review of International Political Economy,* and *New Political Economy.* In 2015, she was awarded the British Academy for the Humanities and Social Sciences' Rising Star Engagement Award.

Richard Murphy is Professor of Practice in International Political Economy, City University. His books include: *Tax Havens: How Globalization Really Works* (Cornell University Press, 2010), *The Courageous State* (Searching Finance, 2011), *Over here and under-taxed* (Vintage, 2013) and *The Joy of Tax* (Transworld, 2015).

Anthony Payne is a Professorial Fellow at the Sheffield Political Economy Research Institute (SPERI) at the University of Sheffield and was the founding co-director of SPERI. He is the author, most recently, of *Civic Capitalism* (Polity, 2015, with Colin Hay), and *The Long Battle for Global Governance* (Routledge, 2016, with Stephen Buzdugan).

He also edited the *Handbook of the International Political Economy of Governance* (Edward Elgar, 2014, with Nicola Phillips).

Jonathan Perraton is Senior Lecturer in Economics at the University of Sheffield. His works include *Global Transformations: Politics, Economics and Culture* (Polity Press, 1999, with David Held, Anthony McGrew and David Goldblatt) *and Where are National Capitalisms Now?* (Palgrave Macmillan, 2004, with Ben Clift).

Nicola Phillips is Professor of Political Economy and Vice-President and Vice-Principal (Education) at King's College London. She has been an editor of both *New Political Economy* and *Review of International Political Economy*. Her books include *The Handbook of the International Political Economy of Governance*, edited with Anthony Payne (Edward Elgar 2014), *Migration in the Global Political Economy* (Lynne Rienner, 2011), and *Development*, with Anthony Payne (Polity Press, 2010).

Helen Thompson is Professor in Political Economy at the University of Cambridge. She is the author of a number of books including *Oil and the Western Economic Crisis* (Palgrave, 2017), *China and the Mortgaging of America* (Palgrave, 2010) and *Might, Right, Prosperity and Consent: Representative Democracy and the International Economy* (Manchester University Press, 2008).

List of Figures

Introduction: The Coming Crisis, the Gathering Storm

Colin Hay and Tom Hunt

Abstract The introductory essay reflects on the sources of disequilibrium and instability in the current context, and makes the case for a form of prospective precautionary thinking as a means of anticipating and protecting ourselves against the coming crisis. It does this through an assessment of the difference between the natural pessimism of the political economist and the natural optimism of mainstream economists. The chapter concludes by developing a short overview of the principal sources of risk and pathology in the UK, European and world economies today, relating these to the argument of the chapters to follow and how they each explore themes of disequilibrium and crisis in our present conjuncture and their potential role in the coming crisis.

Keywords Instability · Disequilibrium · Prediction · Optimism · Pessimism

C. Hay · T. Hunt (✉)
University of Sheffield, Sheffield, UK

C. Hay and T. Hunt (eds.), *The Coming Crisis*, Building a Sustainable Political Economy: SPERI Research & Policy,
DOI 10.1007/978-3-319-63814-0_1

That economists tend to optimism whilst political economists tend to pessimism is something of a truism, albeit one that is all too rarely acknowledged. The reason for this bifurcation in psycho-analytic temperament is simple. Economists tend to think in equilibrium terms, because most of the time the world appears, on the surface at least, to exhibit equilibrium tendencies. They comfort themselves with the following thoughts: (i) that, although disequilibrium scenarios exist, they are inherently difficult to grasp; (ii) that there are a relatively small numbers of cases and such cases as arise are *sui generis*—and need to be understood in their own terms anyway; (iii) that, as such, these exceptional events do not avail themselves readily of mainstream economic modes of analysis (based, as they are, on the elucidation of general laws); and, (iv) given that they are rare anyway and that one has to concentrate on something, having a good appreciation of the 99.9 per cent of the time when things are stable provides at least decent compensation for blanking the difficult 0.1 per cent that remains. Appropriately reassured, they turn a blind eye to disequilibrium even whilst they acknowledge its possibility. In short, they model the world as if it were in a natural and benign condition of equilibrium not 99.9 per cent of the time, but all of the time.

This, in essence, was at the heart of the famous—and, one can only presume, inadvertently brutal—question posed by Queen Elizabeth II to a collection of academic economists at the London School of Economics in November 2008. Why, she asked, had they not seen the 'awful' global financial crisis coming? She didn't receive a direct reply and, given that she asked pretty much the same question again at the Bank of England some 4 years later, was presumably not entirely impressed by the one she eventually did receive. It is not difficult to see why. That response was almost 8 months in the making and came in the form of an open letter, on British Academy letterhead, signed by Tim Besley and Peter Hennessy ostensibly on behalf of the discipline of economics, at least in Britain. It is a remarkable document, not least for the following comment. Yes, it suggested, the premise of the Queen's question was right—the discipline of economics had failed to anticipate the coming crisis.

> But the difficulty was seeing the risk to the system *as a whole* rather than to any specific financial instrument or loan. Risk calculations were most often confined to slices of financial activity, using some of the best mathematical minds in our country and abroad. But they frequently lost sight *of the bigger picture*. (emphasis added)

Put more bluntly, very clever people ('some of the best mathematical minds in our country' no less) doing very clever economics hadn't seen the crisis coming because that wasn't the kind the economics that very clever people do. An holistic analysis of systemic risk was quite beyond them, not because they weren't very good at it, but because they weren't interested in it. This was more of an error of commission, then, than one of omission. British Academy letterhead notwithstanding, this was not a very adequate answer to the Queen's question; and it remains wholly inadequate today. One can only speculate as to what some of the previous incumbents of Britain's royal palaces might have deemed a suitable punishment.

Pessimism of the intellect ...

This is where political economy comes in. It is, of course, quite impossible to imagine a royal audience for a similarly constituted group of academic *political* economists. But what is no less clear is that, in such a hypothetical scenario, the assembled heterodoxy of voices (surely the appropriate collective noun) would not have been found wanting for answers to the same question.

The point is that political economists are almost naturally suspicious of equilibrium and no less naturally fascinated by disequilibrium—the moment, indeed the moments, when it all goes wrong. They justify their suspicion and their fascination very differently from their more mainstream economist counterparts. Yet they typically also appeal to four factors in so doing: (i) crises, though rare, are (or tend to be) genuinely transformatory and cannot simply be dismissed as complicating aberrations; (ii) impressions of equilibrium are typically misleading in that the seeds of crisis, if one is of a mindset to look for them, are invariably present even in contexts which might appear superficially placid and benign (the 99.9 per cent, as it were); (iii) it is almost impossible, without the benefit of hindsight, to differentiate between self-equilibrating processes and cumulatively destabilising processes (such as the inflation of an asset-price bubble)—and it is naïve and dangerous, to mistake the latter for the former; consequently, (iv) one should always be looking for, and alert to the risks of, disequilibrating tendencies even in ostensibly equilibrium scenarios. Political economists, in short, are interested in the relationship between the 0.1 per cent of the time when we acknowledge that things are broken and the 99.9 per cent of the time when we tend to assume they are not.

It is probably clear by now that we, the authors of this collection on 'the coming crisis' are not economists but political economists. Indeed, were we not, we might be rather less inclined to countenance the possibility of a coming crisis. And, in a way, that is the point. For it provides us with the justification for this collection. If we acknowledge that things can go wrong, that it matters (quite a lot, as it happens) when things do go wrong, and that economists (by virtue of their analytically or temperamentally engendered optimism) are not good guides to the likelihood of this, then we need some kind of analytical counterbalance. We need, in short, to reflect not on the last or present crisis, but on the *next*.

That is the task we pose ourselves in this collection. Our aim has been to bring together a distinguished collection of political economic voices and perspectives to reflect on our contemporary condition with an eye to discerning the seeds of future crises in the present.

This might seem like a rather morbid pursuit. And that is perhaps hardly surprising. For we, of course, are the natural pessimists—the Cassandras of modern day economic analysis. But our pessimism now seems more widely shared. For, today, even the economists worry (as do the political elite whom they so frequently advise and the public in the name of whose collective interest this advice is offered).

There are two rather obvious, though very different, interpretations of this. The first is that chastened perhaps by the crisis that engulfed the world economy in 2008 (and their blithe overconfidence that it couldn't and wouldn't happen), economists have become (for now at least) rather less naturally optimistic than they were. The second is that the world has indeed become a rather more dangerous place.

There is surely some truth in both. Here and in the more substantive chapters that follow we are rather more interested in the second. Our aim is to explore the hunch in rather more detail. In seeking to do so, we have invited 14 seasoned SPERI commentators to offer their own distinct perspectives on the pathological symptoms of the present conjuncture—reflecting and speculating on the extent to which the world economy has, indeed, become a more dangerous place (*less* not more stable) in the years since 2008.

What is clear is that there is certainly plenty to concern the pessimist. Domestically, and despite the rhetoric, there has been no 'rebalancing' of the UK economy (Berry and Hay 2016); it remains stubbornly dependent on credit and an overgrown financial sector; growth only seems attainable in and through the periodic pump-priming of the housing market; the housing market, itself, now seems more dependent on overseas demand and that, of course, is now threatened by the depreciation of the currency associated with the huge economic uncertainty precipitated by the vote for 'Brexit'.

At a European level, things are certainly no better. Brexit is again a genuine threat. If Ricardo taught us anything (and here the economists would surely agree) it is that a worsening of the terms of trade between partners hurts both sides. Self-inflicted Brexit, in other words, is unlikely to prove a victimless crime. Yet, Brexit itself feels like part of a wider dynamic—a tipping-point, perhaps, in which European integration could start to give way to a no less protracted but rather different process of European ***disintegration***. What is also clear—indeed, rather more clear (for very little about Brexit is clear at this point)—is that austerity is ravaging Southern Europe and, in combination with the migration crisis, is contributing to the resurgence of a political right across Europe that is likely to accelerate the pace of disintegration, not just economically but also socially.

And on a global stage, there is little to lift the gloom. The period since 2008 is likely to be remembered as one in which the opportunity for global financial market re-regulation and genuine governance was missed. Our banks remain too big or too interconnected or too correlated in their behaviour to be allowed *to fail* and yet too big, too interconnected or too correlated *to bail*. What that means is that a second global financial crisis is certainly no less likely—indeed, for many, it is now inevitable. But what we also know is that the capacity to deal with such a crisis has been significantly eroded by the nature of the public response to the first crisis—to the point where it is no longer clear what the response to a second crisis might now be. And this is before we start to factor in the destabilising consequences of the Trump presidency.

All of this, of course, must be set in the context of the wider global environmental crisis. And when you put the pieces together the coming crisis looks all too real. It is beginning to look like a big one. It is beginning to look like the perfect storm.

… Optimism of the Will

But then, of course, we would say that, wouldn't we? We are, after all, the pessimists … Or are we? For it is not in the end pessimism that drives our search for the sources of future crises in the pathologies of the present conjuncture; that would indeed be a morbid pursuit. Here the tables are turned. For it is the *hope* that such a precautionary reflex might allow us the knowledge to deal with pathologies *before* they precipitate fully fledged crises that leads to our seemingly morbid preoccupation with pathology in the first place. In a sense, then, our precautionary warnings on the potential sources of the coming crisis are borne of an optimism that things can still be done to avert the harm and suffering such a crisis would bring.

This leads us to a final introductory reflection on the nature of crisis itself and the implications for the 'crisis' we typically regard ourselves to have experienced and, perhaps, that to come. For strange though it might seem, our argument is that the more one reflects on this, the less self-evident it is that we have yet to experience the crisis itself. This is a second sense in which it is possible to talk of the 'coming crisis'.

It is clear that the language of crisis has, if anything, been cheapened in recent years. Everything these days, it seems, is in a state of crisis. If so then surely it is not difficult to apply the term to the economic storm unleashed in 2008? If it is not a crisis, then what is? Well, it is certainly bad enough; but in a sense that is precisely the point. For if we return to the (Greek) etymology of the term, we find that a crisis is a *moment of decisive intervention* —medically, the point at which the doctor's intervention proves decisive, one way or the other, in the course of the illness and the life of the patient (Habermas 1975: 1; Hay 1996). If the analogy still holds, we are not yet at that point. For although the patient may well be suffering more than ever and the condition does not seem to be improving this is not because of the failure of any decisive intervention. For none has been attempted. If this is what a crisis is—a moment of decisive intervention—then we have simply yet to get to the moment of crisis. What we have seen is instead the accumulation of a series of largely unresolved contradictions—not that the significance of this should be underemphasised. For in many respects, this is far worse; it would surely be better were we able to talk about this as a moment of decisive intervention.

So what possibility is there of our situation of radical indecision becoming one of decisive intervention? For, perverse though it might seem, the best that we can hope for is a crisis—at least a crisis thus understood.

Here, there are grounds for optimism and pessimism alike.

For the optimist, crises understood as moments of decisive intervention and paradigm shift are rare, though they typically post-date the emergence of the symptoms they ultimately seek to resolve, often by a decade or more. In short, we may be too impatient in expecting the crisis point to have been reached already. This was certainly the case in the 1930s—with the transition to Keynesianism taking at least a further decade from the advent of the great depression; and a similar kind of time lag can arguably be identified in the process of change initiated in the 1970s. It seems that the transitions we now associate with crisis periods take a long time to arise—typically, a decade or more before a new order is fully realised. It is perhaps ever more likely the more the condition remains resistant to the current medicine—medicine, of course, prescribed by doctors trained and versed in the operation of the old paradigm (and its equilibrium thinking).

Yet there is only so much optimism one can draw from such historical analogies. For there can be no guarantee that alternative doctors with alternative medicines will be summoned simply because the patient remains unwell and the condition is not responding to current treatments. Searching for solutions is no guarantee that they are found nor that they are implemented. Above all this is a political problem. For us to respond to the crisis differently, we need a different politics of crisis identification. The problem we face politically is that, to far too great an extent, we have either not been looking for solutions (certainly not for alternatives to the prevailing paradigm) nor, to the extent that we have been looking for solutions, have we been looking in the right places and in the right way. Our hope is that, in the chapters which follow we make a compelling case that we need first to get right what went wrong in order to put it right and to suggest at least some of what 'getting it right' and 'putting it right' might entail. Whether, in the end, we summon the political will to act and to act decisively enough to avert catastrophe is itself politically contingent. And therein lies the real test of whether one is an optimist or a pessimist.

If prediction and diagnosis are two interlinked themes running through this volume, then a third is the difficulty of preventing a coming crisis—'putting things right', as it were, as the effects of what went wrong last time are still accumulating. As Helen Thompson argues, the unpredicted and extreme political shocks of 2016 were arguably the predictable manifestations of the breaking down of the pre-2008 economic, political and geopolitical order. We might not be in Kansas any longer; but we shouldn't perhaps be so surprised that the tornado struck. What we can undoubtedly say with confidence, and with justified pessimism, is that the 'other crisis'—the socio-ecological crisis—is bearing down upon us with alarming speed. Indeed, accelerating global warming, ecosystem degradation and biodiversity loss suggest it may already be here. Martin Craig warns us that the failure to adequately diagnose the socio-ecological crisis risks leaving us with a fundamentally inadequate response—the consequences of which will shape all of our lives—and conceivably the lives of all humans yet to be born too.

Our socio-ecological crisis has been centuries in the making. But other features of our current plight invite comparisons with more recent decades of political and economic crisis. Jeremy Green reflects on the return of stagflation that so alarms policymakers today, and by drawing links between economic stagnation and growing political populism, argues that to resolve the current political crisis of Western democratic capitalism will mean loosening the shackles of market discipline to create a new paradigm of economic governance for our times. New paradigms of governance are similarly called for by Anthony Payne. His chapter addresses the profound and currently intractable problems undermining the current regime of global governance: a regime, he argues, that is simply not currently strong enough to avert a further global economic crisis and that today finds itself caught between new and conflicting 're-globalisation' and 'de-globalisation' political pressures.

Whilst the future form of globalisation and its governance will be highly contested, it is uncontestable that we live in an ever more complexly interdependent world economy in which global supply chains provide work for millions and create enormous (and inequitably distributed) wealth. Yet as Genevieve LeBaron illustrates, current global trends highlight persistently low wages, rising unemployment and severe forms of labour exploitation and suggest that the world is on the verge of a crisis of indecent work. Unequal and unsustainable global wealth creation, and its resultant environmental impacts, are themes further explored by Peter Dauvergne. He argues that much-heralded international environmental agreements fail to confront rising rates of overconsumption, unequal consumption, and wealth inequality. All are central to the 'coming crisis of planetary instability'.

A failure to develop adequate responses to complex problems lies, similarly, at the heart of Nicola Phillips' diagnosis. She examines the European migrant crisis of recent years, its interaction with Southern

Europe's lingering economic crisis, the unequal politics of austerity across the EU and how these issues combine to have profound political, economic and social implications. The failure to resolve deep-seated problems manifests itself in myriad ways; the aforementioned rise in political populism, economic nationalism and growing suspicion of so-called elites and experts, to name just a few.

Jacqueline Best considers this last effect and its corollary of declining legitimacy for political and economic policymakers—a necessary requirement for effective governance that enjoys democratic support. She argues that whilst central banks provided some of the most effective responses to the last crisis, it is unlikely that they will play such a role again. This is in part because their reliance on 'rule-based' monetary policy makes it difficult for them to acquire belatedly the legitimacy and effectiveness needed to fight the next crisis. In fact, their declining legitimacy poses itself a very real threat. The tension between dysfunctional rules and institutional arrangements and intractable political obstacles that prevent them from being replaced is also core to Scott Lavery's chapter on the eurozone. He outlines, how the eurozone continues to be afflicted by a number of profound imbalances. These combine to threaten to undermine its status and position as a viable economic and political unit. As Europe remains trapped by the dysfunctional architecture of its currency, its deep imbalances remain, meaning the continent is likely to be front and centre of any new global economic crisis.

The pressing need to establish a new political and economic settlement for our times is highlighted by various authors. This will both require new institutions to be created and for existing institutions to adapt to new roles. In their chapter, Andrew Baker and Richard Murphy examine how the state and central banks, with their new mandate for 'systemic stabilisation', interact with financial services, and in particular with the complex shadow money system. They argue that any new social contract must tie in financial markets and imbue their actors with a greater sense of their collective social obligation. Nowhere is the centrality of financial services to contemporary Western capitalism more apparent than in the UK. The highly financialised nature of the UK economy is a major factor in understanding the country's weak post-2008 recovery but was also central to the growth model of the pre-2008 'Great Moderation' period. Jonathan Perraton's chapter analyses the UK's prolonged stagnation in productivity, and how the post-crisis return to growth once again appears to be driven by debt-financed private consumption. He questions whether, nearly a decade on from the last crisis, secular stagnation should be seen as the 'new normal' for the UK.

In their different ways all of our authors show how many of the shocks that have shaped, and are shaping, the contemporary global political economy were simply not seen coming. And yet ironically one crisis that has been predicted for decades is the 'coming' economic crisis in China—a crisis that has to date failed to occur. As growth in China is slowing and its burgeoning imbalances are increasingly evident, the crisis drumbeat is getting louder—could the crisis finally be upon us? In answering this question Matthew Bishop charts China's substantial developmental transformation and assesses the likelihood of a crisis and its repercussions for China and the global economic order—if indeed it is to come.

The risks therefore of greater shocks ahead are clear and present. But what we hope to demonstrate with equal clarity is that our ability to avoid them is politically contingent. In concluding, Andrew Gamble reminds us of the centrality of politics. He diagnoses and analyses three phases of the post-2008 crisis period and argues that since 2016 we have entered a fourth phase—a political crisis. As a result, today we are in a radically new *political* environment where the political and economic assumptions of the international market order which has been dominant since 1945 are being challenged as never before.

Storm clouds are undoubtedly gathering. Don't say we didn't warn you.

References

Berry, C., & Hay, C. (2016). The great British "rebalancing" act: The construction and implementation of an economic imperative for exceptional times. *British Journal of Politics and International Relations, 18*(1), 3–25.

Habermas, J. (1975). *Legitimation crisis.* Boston: Beacon Press.

Hay, C. (1996). Narrating crisis: The discursive construction of the 'winter of discontent'. *Sociology, 30*(2), 253–277.

We are Not in Kansas Anymore: Economic and Political Shocks

Helen Thompson

Abstract 2016 was a year of apparent political shocks from Britain's vote to leave the European Union to the election of Donald Trump. Yet whilst most analysis failed to predict these events, they were the clear product of the breakdown of the economic and political world that was in place before 2008. The breakdown of that order has produced two differing sets of consequences in relation to economic and political probability. It has, in conjunction with high-frequency trading, transformed the monetary and financial world making the financial markets the site of black swan events in terms of existing models of financial markets, leaving us in an unknown economic world. By contrast, in politics, there are historical antecedents in past crises to the kind of events that unfolded in 2016.

Keywords Quantitative easing · Zero interest rates · Brexit · Black swan events · US election

---

H. Thompson (✉)
University of Cambridge, Cambridge, UK

C. Hay and T. Hunt (eds.), *The Coming Crisis*, Building a Sustainable Political Economy: SPERI Research & Policy,
DOI 10.1007/978-3-319-63814-0_2

2016 was a year of apparent political shocks. We seemed by the end of the year to have left a political world we understood behind and entered a new one. Put in the language of movie-culture, by the end of 2016, we were no longer in Kansas. The two developments of 2016 that have, of course, attracted most comment have been the Leave win in Britain's referendum on its membership of the European Union (EU) and Donald Trump's victory in the American presidential election. Perhaps just as consequentially, however, 2016 also saw an attempted coup in Turkey during which the US government appeared initially neutral about its outcome and the sight of Russia acting together with Iran and a NATO member in Turkey to the diplomatic exclusion of the United States to negotiate a cease-fire in Syria. Looking at this new political world, we seem to be living in a bewildering, and perhaps terrifying, political time of what could be, and are often, called 'black swan' events; these are events of low probability that are extremely difficult to predict.

Yet even without the benefit of retrospective hindsight, none of these political events in 2016 was in reality as improbable as we may have thought. Put simply, the economic and political world that was in place before 2008 no longer exists, and the events of 2016 are a manifestation of the breakdown of that old economic, political and geopolitical order. In the Middle East, the American failure in Iraq has fundamentally changed the balance of power in the region by strengthening Iran and facilitating a counter-reaction in Sunni majority-states. In this new geopolitical environment, the United States is unable to exercise power in the Middle East in the manner in which it has done since the end of the cold war, and Russia has seized the opportunity to re-enter the region. Within the EU the eurozone crisis has elevated German power and ensured that virtually all further integration will be driven by the need to recreate the institutional basis of monetary union. As a consequence, the EU can no longer function politically as it did a decade ago. Economically, the 2008 crash brought to an end the material, financial, and political foundations of non-inflationary growth in western economies and began a new era of quantitative easing (QE) and zero interest rates policy (ZIRP). The result has been a 40 per cent increase in global debt since 2007 and a radical transformation of the structural conditions of international capital flows, the relative position of creditors and savers, and the fundamental context in which monetary policymakers can judge the likely consequences of their actions.

Certainly, this post-2008 world has produced both volatility and unpredictability, most clearly in the operation of financial markets. Indeed, it may not be hyperbole to suggest that in the wake of QE and ZIRP western financial markets as *markets* with any price discovery function no longer exist. Share and bond markets in particular now have dynamics permeated to the core by expectations of what central banks, and in particular the Federal Reserve Board, will next do. For example, in May 2013, US bond markets threw what became deemed a 'taper tantrum' when Ben Bernanke said that the Fed planned to taper bond purchases under QE3 and in doing so pushed up sharply yields on Treasury bonds. In this new financial world share and bond markets often respond positively to the bad economic news in the real economy because poor data derails further the day when central banks can move back towards anything like a remotely normal monetary policy. As the gyrations of the financial markets over the first twenty four hours of Trump's victory demonstrated, bond and share valuations also react with large and erratic swings to political developments, as investors endeavour to process what political outcomes will do to the likelihood of monetary change.

This increasingly surreal world generated in financial markets by QE and ZIRP has been compounded by the manner in which these markets have simultaneously been recast by high-frequency trading. Correlations between movements in different asset classes from shares to bonds to commodities and between assets in advanced and emerging market economies have become acute since 2010 with a whole range of prices driven by common external developments, not least the pronouncements of the world's central banks, rather than anything particular to the singular fundamentals of each market. Although rising correlation was a predictable feature of periods of high market volatility in the years before 2008, the intensity of the correlation is now levels of magnitude greater than anything seen before. Meanwhile, post-2008 financial markets are producing what would have hitherto been regarded as 'black swan' events, flash crashes and surges of such size that should be extraordinarily low-probability occurrences according to all existing modelling of financial markets. In the context of strongly correlated markets and black swan movements in prices and yields, the risk of a systemic crisis through contagion is considerable and the avoidance thus far of another financial crisis that would dwarf anything that happened in 2008 may be considered but good fortune.

In probability terms, the apparent political shocks of 2016 are not comparable. Although like the financial black swans they would not have occurred in the pre-2008 world, at a historical level they have also been reasonably probable occurrences in the context of the rupture in the economic, political and geopolitical order that has taken place. Put differently, these supposed political black swans are the possible events that, historically, we should expect to occur at least some of the time when underlying stresses in structural fault-lines in political orders break. Certainly, the nature of the qualitative monetary and financial transformation and its fallout in terms of low-probability occurrences has not been without its political consequences, as exhibited in the direct attacks made by both Trump and Bernie Sanders on the Fed's QE programme during the American presidential election. But the monetary and financial metamorphosis since 2008 has not, at least yet, yielded anything that looks so inexplicable in politics.

In part, this relative predictability of the events of 2016 is simply the consequence of the fact that the two disruptive election results were the result of binary events in which by the time of the election one of only two possible outcomes simply had to occur. In the case of the American presidential election, we should not be surprised that one party's candidate was able to construct an electoral college victory with small margin wins in a small number of states in one particular region of the country since the number of states changing hands between the Democrats and Republicans in presidential elections from 2000 has been relatively limited and those changes have determined electoral results. The apparent low-probability event to explain in the American election is how a candidate without previous political experience and little prior attachment to the Republican Party became the candidate of the Republican Party whilst launching an outright assault on the entire political establishment in the US including the Republican Party itself. Nonetheless, even Trump's candidacy is not in probability terms as shocking as it may seem. From Rome onwards, times of crisis in republics and democracies have produced the ascent to power of an outsider member of the dominant oligarchical class, who rises by mobilising the deep discontent of a section of the populace with the ruling elite.

Trump's own relationship to the American oligarchic class, through his celebrity and the material dynamics of campaign finance that the oligarchical components of American democratic politics generate, created his political opportunity to join the race for the Republican nomination.

Once in the contest, the fact of Trump's political inexperience then allowed him to act as an effective whistle blower on the ruling political class' failures to preserve the old economic and political order, not least in regard to the failed, and exorbitantly expensive, wars in the Middle East. When the whole foreign policy-making establishment that had presided over this imperial overshoot lined up against him in the Republican primaries as if nothing had changed in relation to the United States' position in the world, it was in practice relatively easy for Trump to rally a large enough constituency of voters by pointing out that the US could no longer afford to play unsuccessfully at being the world's policeman.

In the case of British membership of the EU, a Leave result in a binary referendum was an even higher probability event from the outset. What requires more explanation in accounting for Brexit is why David Cameron first gambled on such a binary referendum to determine whether Britain would remain inside the EU when that was an outcome to which he was strongly committed, and then was unable to persuade other EU leaders, and in particular Angela Merkel, that ongoing British membership was worth significant concessions. Here again, the elucidation of these outcomes lies in the breakdown of the pre-2008 economic and political order, this time in relation to the EU. The eurozone crisis put massive pressure on the foundations of Britain's membership of the EU. In general terms, it politicised the position of London as the eurozone's offshore financial centre, it created the need for further integration of which Britain as a non-participant in the eurozone would have no part, it turned Britain into a joint employer of last resort with Germany for the periphery of the eurozone, and it magnified the differences in approach to monetary and financial matters between Britain and the other non-euro member states.

Cameron walked the path to his referendum promise in 2013 because he was unable to find an alternative way out of the political pressures these dynamics created, and he secured little in the renegotiations from Merkel in 2016 because under the conditions of the eurozone crisis British membership mattered significantly less to the future of the EU than it had done before 2008. Put more schematically, the pre-2008 centre of Britain's membership of the EU no longer held. If there was no necessary reason why any British Prime Minister had to confront that reality, or in Cameron's case to conclude that he could change it by reconstructing the domestic foundations of Britain's place in the EU through renewing democratic consent to the basic principle of membership, there was also no path back to a world in which that centre existed. Seen from history, the departure of a large state from a confederal or federal union in which it had long been in a political minority at a time when a crisis exposed the limits of that state's political influence within the union would appear a not unlikely event at all.

The world in which Britain is leaving the EU and a political neophyte who declared rhetorical war on the American establishment is President of the United States is both unrecognisable in relation to the pre-2008 order and could have been predicted as a possible consequence of the kind of disjuncture that 2008 represents. The conjunction of developments that brought those elections to the particular binary choice at issue, which in both cases pitted attempted continuity against disruption, arose in the context of a disorder that had by definition to advantage disruption. Of course, structural advantages do not determine in politics and in particular they do not decide binary elections in which day-to-day events are highly charged and fast-moving and voter turnout is variable. Either election could have produced a different outcome if a number of contingencies had been otherwise. This is particularly true in the case of the American election where Trump's path to an electoral college victory turned on extremely small margins. Nonetheless, historical experience of economic, political and geopolitical crises and the disorder they let loose tells us that radical political change often ensues under the kind of conditions now in play, especially when, as in the United States, economic and geopolitical crises occur simultaneously.

There is a coming economic and political crisis. What history cannot predict with anything like such clarity is the future economic and political outcomes that the monetary transformation wrought by QE and ZIRP will eventually yield. There has simply been nothing in human history that looks like this monetary experiment in which central banks have created from nothing a massive volume of new money to service and expand debt whilst permeating in doing so the whole nature of financial markets. In this respect, we have indeed left Kansas behind and are living in an unknown Oz.

On 'the Other Crisis': Diagnosing the Socio-Ecological Crisis

Martin Craig

Abstract A 'socio-ecological' crisis is unfolding between our societies and the ecologies of which they are a part. Exploring some of its key dimensions reveals that its precise nature is not as clear as is sometimes assumed. A little acknowledged, but extremely significant, debate over the 'diagnosis' of the crisis exists, which in turn reflects a long-running dispute between 'under-consumptionist' and 'over-accumulationist' currents in crisis theory. The debate points us to the very different and incompatible 'prescriptions' that the two currents offer for how to avoid the worst impacts of the crisis. Diagnosing the socio-ecological crisis is therefore central to understanding 'the coming crisis'.

Keywords Socio-ecological · Consumption · Accumulation · Diagnosis Capitalism

These are anxious times for political economists, as this book certainly attests. A crisis is coming (if, indeed, it is not already upon us), but what

M. Craig (✉)
SPERI, University of Sheffield, Sheffield, UK

C. Hay and T. Hunt (eds.), *The Coming Crisis*, Building a Sustainable Political Economy: SPERI Research & Policy, DOI 10.1007/978-3-319-63814-0_3

kind of crisis is it going to be this time? And how does it relate to the 'other crisis'—that unfolding ecological crisis of which we are all aware, but which political economists are so often reticent to discuss?

Drawing upon ideas from the overlapping fields of political economy and political ecology, I make three points by way of a brief engagement with these questions. The first is that the coming crisis and the 'other crisis' are one and the same. This crisis is neither a social nor an ecological crisis, but a 'socio-ecological' crisis, arising between the complex web of human relations that we call 'society' and the complex web of ecological relations of which society is an irreducible part, with far-reaching implications for both. Second, the precise 'diagnosis' of this socio-ecological crisis is presently uncertain (Craig 2017): a little-acknowledged debate exists here, and some interpretations suggest a more intractable threat than others. These interpretations correspond to a venerable but unresolved dispute between political economists rooted in 'underconsumptionist' and 'over-accumulationist' schools of crisis theory (cf. Stockhammer 2013; Harvey 2010). Finally, I make the point that these diagnoses of the socio-ecological crisis imply rather different prescriptions, making the quality of our diagnosis all the more important, lest ineffective or counterproductive prescriptions be drawn.

Conceptualising Capitalism's Socio-Ecological Crisis

Discussion of 'ecological crisis' often centres on what contemporary capitalist political economies 'do to' nature as they develop and grow—think pollution, biodiversity loss, global warming, and so on. But, as Jason Moore (2015) has recently observed, an equally important question asks what nature '*does for*' capitalism—how it provides the conditions for successful capital accumulation, and the growth, employment and development upon which its stability and popular legitimacy as a form of political-economic organisation depends.

Capitalism's troubles—as other contributors to this book rightly observe—begin when such conditions are no longer available. As political economists, we are accustomed to thinking of these conditions in terms of social institutions arranged across the local, national or international levels. Some 'models of capitalism', the argument goes, are better able than others to facilitate the kind of long-term investment, aggregate demand and rates of worker income that a capitalist political economy requires in order to grow in an inclusive manner (Hay and Payne 2015).

True enough, yet at a deeper level, it is also true that a capitalist economy requires certain ecological conditions to be in place too. As well as a habitable planet (which, after all, is a precondition of any kind of society), capitalist accumulation and growth require access to natural resources sourced from the material ecological context of which capitalist societies are a part. These resources include labour and food, as well as the raw materials and energy which we are more accustomed to thinking of in these terms. Moreover, the supply of these basic commodities must be both plentiful and cheap if production costs are to be controlled and capital accumulated (Moore 2015). If the cost of these basic commodities were to rise in a concerted and prolonged manner (perhaps due to excess demand, or growing expense/declining productivity in their supply resulting from environmental degradation or depletion), then the result would be pressures on system-wide profits. This, in turn, would have huge implications for future economic growth, employment, equality, financialisation, and international relations—in short, it would intensify all of the political-economic crisis tendencies pointed to elsewhere in this book.

In this sense, 'ecological crisis' can be a crisis *for* capitalist societies. Yet the idea of 'socio-ecological crisis' takes us a step further, suggesting that ecological crisis and social crisis are inextricably linked. The reason, simply stated, is that production in capitalist political economies is driven towards continual expansion, creating ever-expanding demands for the basic commodities outlined above (see Blauwhof 2012, for a decisive rebuttal of the possibility of 'steady-state capitalism'). All else being equal, this leads to ever-increasing pressures on the ecological relationships from which these basic commodities are drawn, rendering resources scarce and expensive through depletion, excess demand and environmental degradation. The result is that capitalist political economies face increasing difficulties maintaining the ecological conditions for capitalist accumulation and growth.

Yet to date, all else has not been equal. Historically, capitalism's tendency to exhaust its own ecological conditions have been offset by the incorporation of new 'frontiers' (new sources and new forms of raw materials, energy, food and labour) into the capitalist world political economy. A growing historical literature in the field of political ecology makes this case by conceptualising and analysing capitalism's evolution from its late medieval European origins to its contemporary global form (Smith 1984; O'Connor 1998; Moore 2015). This historical work notes the cultural, institutional, technological and scientific changes in capitalist societies that have allowed the ecological conditions of capitalist accumulation and growth to be remade in expanded quantities at different moments in history, even as the scale of capitalist production continually expands. Some such accounts (e.g. Moore 2015) have sought to periodise these changing configurations into distinct 'world ecological regimes', broadly corresponding to different 'regimes of accumulation' that have structured the world political economy over the centuries (on which see Arrighi 1994).

What this historical work sheds light on are the paradoxical implications of capitalist growth: the way in which the ecological conditions for capitalist accumulation and growth have been secured at one historical moment have themselves proven crisis-inducing at later moments. One paradigmatic example concerns the movement from the predominantly biomass-based energy system in late medieval times to a primary fossil fuel-based system now. This transition freed capitalist production from the constraints of available biomass, but the resulting net increases in atmospheric CO_2 now underpin global warming. Another example concerns the high-productivity agricultural practices of the mid-twentieth century 'green revolution', through which global food supplies were able to keep pace with rising populations and rising per-capita food consumption: the very same practices are now a major factor in biodiversity loss (Rockström et al. 2009). Both global warming and biodiversity loss represent the overshoot of planetary 'life support systems' and a threat to the survival of contemporary societies in a recognisable form (ibid). Moreover, both fossil fuels and green revolution agricultural techniques appear now to be stagnating as means of delivering low-cost energy and food, as the increasing capital intensity of both sectors demonstrates. In this sense, a socio-ecological crisis can also be seen as a crisis *of* capitalist society, albeit one that has been continually (but temporarily) displaced in time and space.

Diagnosing Capitalism's Socio-Ecological Crisis

All of this may inspire despair or optimism, depending upon how one diagnoses capitalism's present encounter with the socio-ecological crisis. For whilst, there is a little ambiguity that capitalist societies are again confronting the symptoms of a socio-ecological crisis, there is much less certainty as to its precise nature, and thus on how to address it.

There are at least two possible interpretations, one rather less tractable than the other. The more optimistic view characterises the crisis as simply the latest iteration of the cycle traced by historical political ecologists noted above: it is a moment at which the existing institutions of contemporary capitalist societies are unable to secure the ecological conditions for capitalist accumulation and growth, but in which new institutions could emerge that would allow them to do so (Green New Deal Group 2008). Proponents of this interpretation argue that new materials and new approaches to energy and food production already exist that could simultaneously mitigate the degradation of the human habitat *and* keep the costs of these basic commodities low—the challenge lies in devising effective political strategies to navigate and ultimately displace those social forces that favour the crisis-ridden status-quo (Newell and Paterson 2010). Yet from this perspective, there is no reason in principle why this feat should not be accomplished: capitalism can survive this crisis of its own making, as it has done in similar such instances in the past.

The task is no small one, especially in the staid economic circumstances of the post-2008 context. Yet some optimistic takes on this perspective go even further, envisioning not only the flourishing of capitalism per se, but also the possibility of a much more equitable model of capitalism *through* the construction of new 'green growth models'. Often referred to as 'the green new deal', this approach seeks to generate a new cycle of employment-rich capital accumulation through the production and deployment of technologies and infrastructure that will greatly reduce the impact of growth upon the planet, whilst using the proceeds of this accumulation in a variety of socially useful ways (Green New Deal Group 2008).

Proponents are also quick to point out that such a prescription requires a substantial change to the political-economic status quo. They advocate the forging of new national and international economic policy consensuses that depart from both the neoliberalising thrust of contemporary capitalist restructuring, and reverse the resultant financialisation that has both underpinned and destabilised recent global growth. In this view, capitalism's present economic, social and environmental pathologies are caused by neoliberalisation, which has eroded the kinds of public institutions able to sustain aggregate demand and direct economic activity towards socially useful and non-ecocidal outcomes. This perspective, therefore, represents an ecological and progressive inflection on a venerable tradition of political-economic thought stemming back to the work of John Maynard Keynes, somewhat inelegantly termed 'underconsumptionism' in contemporary crisis theory, due to its emphasis on inadequate levels of aggregate demand (Keynes 1936).

A second interpretation is less optimistic about the scope for capitalism to be re-established on this equitable and more ecologically attuned footing. According to this perspective, there are now insurmountable barriers to re-establishing the ecological conditions for capitalist accumulation and growth. This is due to the unavailability of a plausible technological, scientific, cultural and/or institutional reconfiguration able to simultaneously release expanding flows of low-cost basic commodities whilst also addressing the degradation of the planet's life support capacity. In this view, the present crisis does not represent a cyclical moment in the long socio-ecological history of capitalism, but a terminal crisis of capitalism, to which the only effective prescription is a completely new model of political-economic organisation (O'Connor 1998; Moore 2015; Foster et al. 2010; Harvey 2014).

In a prominent statement of this perspective, Jason Moore highlights trends in food prices and agricultural productivity, holding these as evidence that the techniques comprising the so-called green revolution have now peaked as a method of expanding food production without a corresponding increase in the unit price of food (Moore 2015). He cites the global stagnation in the yield growth of staple crops since the 1970s, the thus-far lacklustre impact of biotechnology on the yield potential of cropland, and the erosion of diminishing productivity gains through the invasive species and pesticide resistant 'super weeds' that have emerged as a result of contemporary agricultural techniques and supply chains.

The end of 'cheap food' poses a fundamental challenge to capitalist political economies. Food prices dictate the point of subsistence, providing an absolute floor beneath which wages cannot be pushed without provoking starvation and social instability. In a world in which agricultural productivity fails to keep pace with population growth and neoliberal capitalism's demand for ever-cheaper labour, the political economies constituting the capitalist core would either have to see wages forced below subsistence level (massively intensifying distributional conflict and social instability), or appropriate supplies of food that would otherwise be consumed in the periphery (in effect exporting the crisis, and provoking who knows what manner of conflicts in the present uneasy global context).

Of course, there are various short-term 'fixes' that might be proposed to secure a rising supply of cheap food even in the absence of a revolution in agricultural productivity (insects, the UN Food and Agriculture Organisation tells us, may soon be on the menu; FAO 2013). However, more complex variants of the same argument can be constructed with reference to a range of commodity and fuel price data. All point to yet more uncertainties concerning contemporary capitalism's future access to flows of low-cost basic commodities upon which growth depends.

This interpretation of socio-ecological crisis, like the first, represents an ecological inflection on an established current in political-economic crisis theory, but one of a rather more pessimistic nature. It depicts a form of 'over-accumulation' crisis, characterised by a lack of profitable outlets for accumulated capital (Harvey 2010). The lack of investment opportunities in this instance arises from the increasing difficulties in accessing basic commodities at a low enough price to ensure system-wide profits. Interestingly, the interpretation of financialisation and neoliberalisation that stem from this perspective are the opposite to that of the under-consumptionist narrative: the neoliberal restructuring of capitalist political economies to facilitate wage repression and speculative financial investment is a *reflection* of a lack of profitable outlets for capital in productive activities, rather than the cause of economic stagnation.

The result is an incompatibility between the prescriptions that follow from the two diagnoses: from an over-accumulationist perspective, the kinds of green new deals advocated by under-consumptionists are impossible by dint of their unprofitability—nothing but the transition to a post-capitalist political economy will suffice if socio-ecological crisis is to be resolved.

Reasons to Be…Anxious

The debate between under-consumptionist and over-accumulationist crisis theorists that underpins that over socio-ecological crisis diagnosis has played out among political economists for well over a century. My intention here is not to endorse either interpretation, for the ecological dimension in the debate is relatively new, and the empirical work supporting of either remains as yet underdeveloped (Craig 2017). Rather, my intention is to highlight the ongoing relevance of this debate and its uncertain implications as we apprehend the coming crisis. Despite the perennial (if wholly justified) gloom that characterises our field, the work of those political economists who have addressed the 'other crisis' is often infused with a pervasive optimism about the capacity of capitalist societies to adapt in crisis-displacing ways. Yet, the work of historical political ecologists sensitises us to the potential limits of institutional and technological fixes to capitalism's present socio-ecological crisis. Few acknowledge that this important debate over socio-ecological crisis diagnosis exists, yet a failure of diagnosis risks leaving us with a fundamentally inadequate response—and that, surely, is a reason to be anxious.

References

Arrighi, G. (1994). *The long twentieth century*. London: Verso.

Blauwhof, F. B. (2012). Overcoming accumulation: Is a capitalist steady-state economy possible? *Ecological Economics, 84*, 254–261.

Craig, M. P. A. (2017). *Ecological political economy and the socio-ecological crisis*. Basingstoke: Palgrave Macmillan.

FAO. (2013). *Edible insects: Future prospects for food and feed security*. Rome: FAO.

Foster, J. B., Clarke, B., & York, R. (2010). *The ecological rift: Capitalism's war on the earth*. New York: Monthly Review Press.

Green New Deal Group. (2008). *A green new deal*. London: Green New Deal Group.

Harvey, D. (2010). *The enigma of capital*. Oxford: Oxford University Press.

Harvey, D. (2014). *Seventeen contradictions and the end of capitalism*. London: Profile Books.

Hay, C., & Payne, A. J. (2015). *Civic capitalism*. Cambridge: Polity Press.

Keynes, J. M. (1936). *The general theory of employment, interest and money*. London: Macmillan.

Moore, J. W. (2015). *Capitalism in the web of life*. London: Verso.

Newell, P., & Paterson, M. (2010). *Climate capitalism*. Cambridge: Cambridge University Press.

O'Connor, J. (1998). *Natural causes: Essays in ecological Marxism*. New York: Guilford Press.

Rockström, J., et al. (2009). A safe operating space for humanity. *Nature, 461*, 472–475.

Smith, N. (1984). *Uneven development: Nature, capital and the production of space*. Oxford: Basil Blackwell.

Stockhammer, E. (2013). Financialisation, income distribution and the crisis. In S. Fadda & P. Tridico (Eds.), *Financial crisis, labour markets and institutions* (pp. 98–120). Abingdon: Routledge.

Stagflation and the Shackles of Market Discipline

Jeremy Green

Abstract Western capitalism has seen a return to stagflation since the global financial crisis. Through analysis of the causes and consequences of the stagflationary crisis, and by assessing the theory of secular stagnation, it is clear that major changes in economic policy are required to restore sustainable growth and rebuild a legitimate basis for democratic capitalism. Technical proposals to enhance demand through monetary policy innovation exist, yet arguably the real obstacles to breaking the stagflationary impasse are political and ideological, rather than simply technical. The pre-eminence of market discipline, as the organising principle of political economy over recent decades, must be surpassed. By drawing links between the stagnation of Western capitalism and the rise of populism, the chapter argues that the current political crisis of liberal democratic capitalism can only be averted by harnessing democratic mobilisation to create a new paradigm of economic governance.

Keywords Stagflation · Market discipline · Secular stagnation
Western capitalism · Democracy

J. Green (✉)
Department of Politics and International Studies,
University of Cambridge, Cambridge, UK

C. Hay and T. Hunt (eds.), *The Coming Crisis*, Building
a Sustainable Political Economy: SPERI Research & Policy,
DOI 10.1007/978-3-319-63814-0_4

The darkest hours of the Global Financial Crisis conjured haunting visions of the worst-case scenario: that we would enter a period of sustained financial paralysis, massive unemployment and the complete breakdown of the liberal economic order—in short, a new Great Depression. Now, almost a decade from the outbreak of the crisis, the worst of those fears have been allayed. Growth has returned, albeit stuttering, uneven (with the United States leading, Britain and the eurozone lagging) and modest, and employment levels have made some recovery.

Use of the term 'recovery' must, though, be strictly qualified here. It is too strong a word to describe the anaemic economic performance of the post-crisis era. Certainly, the worst excesses of the 1930s were avoided, thanks, in part, to the lessons learned from the chastening experience of the Great Depression. This time, the US Federal Reserve did not raise interest rates and allow banks to fail en masse. Instead, alongside the Bank of England, it unleashed the full force of its monetary ammunition to prop up financial markets and solidify banks' balance sheets. Total collapse of the banking system was avoided. But to the extent that important monetary lessons were learned from the crisis of the 1930s, the reverse is true for the fiscal lessons that should have been derived from that era. Keynesian counsel for counter-cyclical spending and public investment was conveniently forgotten. In macroeconomic terms, this was a case of selective memory. Once the financial sector had been shored up, governments chose to cleave fervently to the supposed virtues of fiscal austerity. The fiscal amnesia afflicting Western governments has had grave consequences, threatening not only the health of the economy but the very vitality of liberal democracy too.

The Return of Stagflation

This approach to governing the crisis could only take us so far. It is now clear, with the benefit of hindsight, that what could not be avoided was a new slow-burning crisis of stagflationary capitalism in the West: a combination of slow/stagnant growth, unemployment and increasing underemployment, deflation (price decreases) or disinflation (a declining rate of inflation). The 'new normal' of Zero Lower Bound and (more recently) negative interest rates alongside sustained monetary easing has been allied with underwhelming growth performance (IMF 2016). Collapse was avoided but the crisis came at the price of sustained and apparently intractable stagnation.

To contextualise this second stagflationary regime of Western capitalism, the one that began from 2009, we need to return to another crisis period within global capitalism. For what began as a banking crisis that echoed the Great Depression has become one that holds many instructive parallels with that of the 1970s. The stagflationary crisis of the 1970s, driven by oil price increases, rising wages and declining profitability, heralded an unprecedented combination of high inflation, high unemployment and low growth. In doing so, it disrupted reigning Keynesian ideas, which struggled to find the correct policy response. The present macroeconomic malaise shares features with that of the 1970s: low growth and prices stubbornly fixed around an extreme point of the continuum.

But whereas the first stagflationary regime that unfolded in the 1970s was marked by the problem of high and sustained inflation, it is sustained global deflation and disinflation that now threaten the stability of the global political economy. Consumer price inflation in the Western political economies was 0.3 per cent in 2015, the lowest level since the crisis. By 2016, it had risen only to 0.5 per cent. Eurozone economies, trapped within a gold standard-style logic of competitive internal devaluation and austerity, experienced outright deflation (IMF 2016). Stubbornly low inflation runs the risk of producing a deflationary cycle. As firms and households anticipate lower future inflation they suspend investment and consumption. This produces a situation in which weak demand and deflation reinforce each other, arresting growth prospects.

The difference between these two stagflationary phases of post-war capitalism is partly a consequence of the divergent trajectories of oil prices within these two periods. During the first stagflationary regime, OPEC oil price hikes and geopolitical dynamics pushed up prices. Today, the profound collapse of prices in the global oil market since 2014 has had deflationary effects (Obstfeld et al. 2016). This flies in the face of conventional economic logic, which suggests that the decline in the price of a key commodity such as oil should boost consumer demand and lower production costs in oil-importing markets, boosting growth. This tendency should be reinforced by ultra-low interest rates. So why do the Western political economies remain mired in stagnation?

Secular Stagnation

The failure of lower oil prices to fuel growth owes much to the lack of demand within Western political economies. As Larry Summers (2016) has outlined in his 'secular stagnation thesis'—the efficacy of loose monetary policy as a demand stimulus appears to have reached its limits. No matter how far central banks push down the cost of money to stimulate recovery, the demand for borrowing and investment remains stubbornly low. Nearly ten years after the crisis, interest rates remain near the Zero Lower Bound and economic performance lags previous recoveries.

Secular stagnation is another case of going back to the future: working through historical precedent to help capture the peculiarities of our own time. The thesis was first propounded by another American economist, Alvin Hansen, in the wake of the Great Depression. Attempting to understand the problem of slow economic growth and depressed conditions that had blighted economies throughout the 1930s, Hansen (1939) challenged the traditional explanatory focus upon short-term fluctuations in the business cycle. Instead, he pointed to the role of deeper, underlying, structural factors that accounted for slow growth, depressed demand and slack in the labour market. Chief among these factors was the slowing of population growth. As population growth declined, the demand for economic output would also shift. Rather than requiring the capital-intensive development of new housing stock for a growing population, demand would adjust to a much less capital-intensive emphasis upon personal service provision for the elderly. Add this to a decline in the requirement for capital-intensive new technologies and the exhaustion of new territorial horizons for market expansion, and you arrive at a situation of sustained low-investment, low consumption and low growth. In other words, 'secular stagnation'.

In Summers' (2016) updated thesis, the secular stagnation framework is refitted to several contemporary structural trends that are combining to produce a new era of stagnation. The emergence of cash-rich high-tech firms that reap large returns from relatively low capital expenditure (companies such as Apple and Google), slowing demographic growth, and increased income inequality, have led to a long-term structural decline in the demand for investment within the West. The thesis of depressed demand lying at the heart of the Western post-crisis malaise is now widely supported. It suggests that monetary policy alone will not be enough to exit economic stagnation. We need to rediscover the role of fiscal policy as an engine of demand and an agent of progressive redistribution.

Monetary Innovation and the Logic of Market Discipline

If these ideas have gained traction, then why are we not moving to stimulate demand through other measures? It is not for want of ideas and policy proposals. Financial technocrats have begun to consider such a move. Adair Turner (2015), former chief of the UK's Financial Services Authority, has called for 'monetary financing': using central bank money creation to directly stimulate demand through enabling tax cuts or increased public expenditure to finance fiscal deficits without incurring new debt obligations.

An even more direct measure would be to credit the bank accounts of ordinary workers. Turner recommends targeting poorer citizens, as they have a higher propensity to consume rather than save. According to Turner (2015), these policies provide a more direct and effective stimulus than the financial-market mediated and regressive distributional consequences of Quantitative Easing, which has differentially benefited wealthy asset holders through inflating asset prices while failing to lift long-term growth. There is much to commend in Turner's bold proposal. But such measures are unlikely to be undertaken without substantial political change. This is because, at its heart, the question of how to escape the new stagflation involves a crucial political dimension: discipline.

Governing principles of political economy have elevated a certain kind of discipline, 'market discipline', as the central credo of institutional restructuring and macroeconomic policymaking. This has manifested itself in a fixation with fiscal rules, inflation targets, the retrenchment of welfare and rolling out of workfare, and the identification of trade unions as the obstacle to the proper functioning of the market. The overarching ambition of these disciplinary commitments is to maintain the pre-eminence of market forces as the dominant distributive mechanism within society. The genies of extended statist economic intervention and a powerful well-employed workforce, which reared their heads during the crisis of Keynesianism in the 1970s, must not be let back out of the bottle.

The contradictions of the Keynesian commitment to full employment had long been understood. As the threat of unemployment and underemployment subsided, workers would feel empowered to push for higher wages, corroding the profit margins of business and diminishing private control over investment decisions (Kalecki 1943). Economic stagnation brought these tensions to the fore during the 1970s. A reassertion of market discipline over labour and government was the triumphant response.

In practice, discipline has only ever been applied inconsistently, with the poorest and weakest most likely to be targeted while the powerful and systemically significant, as the bank bailouts demonstrated, are likely to be spared the full force of market competition. The selectivity of discipline has been demonstrated in the relationship between states too. Germany broke the rules of the EU's Stability and Growth Pact in the years before the crisis, but nevertheless, insists that Greece must now accept the harsh medicine of austerity and respect European rules.

Monetary financing entails two significant relaxations of discipline that jar fundamentally with the reigning political economy. First, it would free up central banks to create money to finance expenditure and consumption directly, straining their mandate as bastions of the discipline of sound money and further undermining the imagined firewall between a monetary and fiscal policy that maintains their formal institutional independence. Second, it would (if the policy of an electronic 'helicopter drop' direct to citizens' bank accounts was enacted) lead to the substantial crediting of workers' bank balances regardless of their market-based employment activity, thus cutting against the grain of market discipline and welfare retrenchment, as well as re-legitimating an expanded and potentially progressively redistributive state role in macroeconomic management. Although well-intentioned, calls for a 'People's Quantitative Easing' underestimate the profound political obstacles to achieving an expansionary fiscal response to the stagflationary crisis. After years of ideological onslaught against the viability of public investment and public service provision, a wholesale endorsement of fiscal expansion is unlikely. Governments throughout the West, as well as opposition political parties, have staked their reputations on their commitment to an agenda of fiscal austerity and increasing labour market flexibility, and moreover on the propriety of austerity (Blyth 2016). The obstacles to these proposals are political and ideological, not technical.

Absent a coherent intellectual and political challenge to neoliberal dominance, the current stagflationary inertia, which leaves us mired in a fragile, high-debt, low-wage, low-investment and low-growth regime, looks set to endure. Monetary loosening and correspondent asset price inflation are acceptable. But the loosening of market discipline upon ordinary workers and the turn to monetary financing of fiscal expenditure is not.

The Threat to Liberal Democracy

What are the implications of all of this for the likelihood of a new crisis? If the stagnation of Western political economies continues and, indeed, if the likely lurch back into recession occurs, then the already fragile foundations of post-war democratic capitalism in the West will be put under even further strain (Streeck 2014). Continued deflation and disinflation have dangerous economic and political consequences. Much of the deflationary and disinflationary tendencies are accounted for by the continuing depression of wages, particularly in the eurozone. With wage growth stalled, the real burden of debt will increase, while stagnating growth undermines the capacity for its repayment. The current phase of stagflation pits debtors and creditors against each other. Creditors' benefit from the rising real value of their loan assets and interest receipts, but debtors' losses tend to outweigh the potential for creditors' increased spending. This leads to an overall decline in economic activity (Fisher 1933). This is particularly true in an era of financialised capitalism that channels savings into speculative financial assets rather than investment in the real economy. With wages already pushed down to extremely low levels, firms are likely to respond to future economic shocks by laying off workers, rather than pushing for downward wage adjustment (IMF 2016).

Most importantly, stagflation provides fertile ground within which the seeds of populism can take root and grow. The sense of declining living standards and fading economic glory provided the mood music for both the election of Donald Trump in the USA, and the vote for Brexit in the UK. Across Europe, economic malaise is fuelling the rise of anti-European and anti-globalisation sentiment. Authoritarian and xenophobic populism will likely gain further traction. But so too might more progressive forms of populist mobilisation. Populist insurgency, despite its many risks, unleashes energies of mass participation and re-engagement with politics. Those energies are a much-needed antidote to voter apathy and a prerequisite for recalibrating capitalism on a legitimate basis. But the danger is that revived democratic energies can produce dangerous outcomes when they are shaped by a context of economic dismay, despair and detachment from the wider world. Here a central political lesson of the 1920s and 1930s is worth recalling: democratic societies can only absorb so much economic strain under the pressures of austerity and economic stagnation. Persistently low wages and a mood of economic malaise have weakened the enthusiasm for liberal democratic capitalism and undermined support for cosmopolitan values.

The political crisis of Western capitalism has already arrived. But there are still grounds for optimism. Contemporary populism is, at heart, paradoxical: it is both a threat to liberal democracy and the best hope for the revival of democratic political energy. Addressing the economic and social challenge of stagflation will be crucial to containing populism's dangers and harnessing political revival for positive ends. What will this entail? It will mean loosening the political and intellectual shackles of market discipline and advancing a political economy that re-energises the intervention of the state to secure investment, redistribution and sustainable growth. In the wake of previous crises, new paradigms of political economy emerged, gradually and through political struggle, to steer the policy response. First, Keynesianism emerged triumphant from the ashes of the Great Depression and the ruins of war. Then, neoliberal ascendancy was incubated during the crisis of the 1970s. A new paradigm for our own times must both recognise and rationalise the limits of the market and ally this with a reassertion of democratic, egalitarian and green values. If we are unable to do this, then the stagflationary crisis of Western capitalism is likely to continue.

References

Blyth, M. (2016). Policies to overcome stagnation: The crisis, and the possible futures, of all things euro. *European Journal of Economics and Economic Policies: Intervention, 13*(2), 215–228.

Eichengreen, B., & Temin, P. (2000). The gold standard and the great depression. *Contemporary European History, 9*(2), 183–207.

Fisher, I. (1933). The debt-deflation theory of great depressions. *Econometrica: Journal of the Econometric Society, 1*(4), 337–357.

Hansen, A. H. (1939). Economic progress and declining population growth. *The American Economic Review, 29*(1), 1–15.

IMF. (2016, October). Subdued demand: Symptoms and remedies. *World Economic Outlook.* Washington: IMF. Available from: http://www.imf.org/external/pubs/ft/weo/2016/02/. Accessed 20 Feb 2017.

Kalecki, M. (1943). Political aspects of full employment. *The Political Quarterly, 14*(4), 322–330.

Obstfeld, M., et al. (2016, March 24). Oil prices and the global economy: It's complicated. *IMFdirect.* Available from: https://blog-imfdirect.imf.org/2016/03/24/oil-prices-and-the-global-economy-its-complicated/. Accessed 3 Apr 2016.

Streeck, W. (2014). *Buying time: The delayed crisis of democratic capitalism.* London: Verso Books.

Summers, L. H. (2016). The age of secular stagnation: What it is and what to do about it. *Foreign Affairs, 95*(2), 2–9.

Turner, A. (2015, November). *The case for monetary finance–An essentially political issue.* IMF, 16th Jacques Polak Annual Research Conference. Available from: https://www.imf.org/external/np/res/seminars/2015/arc/pdf/adair.pdf. Accessed 14 Apr 2016.

Can Global Governance Prevent the Coming Crisis?

Anthony Payne

Abstract Serious problems undermine the current regime of global governance and create a significant 'global governance deficit'. They have not been solved by the G20, the self-styled steering committee of global economic governance, even though the latest G20 summit in 2016, held for the first time in China, did promote a new 'Hangzhou Consensus' that constituted at the very least a marker on behalf of the cause of more inclusive growth. Actually existing global governance is simply not strong enough to avert a further global economic crisis by its means. It is also now caught between conflicting 'reglobalisation' and 'deglobalisation' political pressures.

Keywords Global governance · Globalisation · G20 · Institutions Global economy

Nobody can say that the major institutions of global governance haven't noticed the possibility that a further global economic crisis might be

---

A. Payne (✉)
SPERI, University of Sheffield, Sheffield, UK

C. Hay and T. Hunt (eds.), *The Coming Crisis*, Building a Sustainable Political Economy: SPERI Research & Policy, DOI 10.1007/978-3-319-63814-0_5

brewing. Indeed, they have been speaking out in fear of that prospect for some while now.

The World Bank warned as long ago as January 2016 that a 'perfect storm' could be building in the global political economy. It was worried by the potential combination of a simultaneous slowing of economic activity across the BRIC (Brazil, India, China, Russia) countries and what it euphemistically termed 'financial market stress'. The International Monetary Fund (IMF) followed the same line. In a speech delivered in Germany, in April 2016, its Managing Director, Christine Lagarde, spoke of her fear that the global economy had lost its growth momentum and was stuck in the 'new mediocre'. 'We are on alert', she said, but 'not [yet] alarm'.

As if to complete the set, a further expression of anxiety came in June last year from the Organisation for Economic Cooperation and Development (OECD), the Paris-based body which has become increasingly important in shaping and delivering global economic governance. Its chief economist, Catherine Mann, introduced the publication of the OECD's latest economic outlook by identifying the emergence of 'a self-fulfilling low-growth trap' over the eight years since the financial crisis broke. The longer this remained the case, she observed, the more difficult it would be to break the 'negative feedback loops'. The language was lumpy and technical, but her message was clear. The resulting risk was that a 'negative shock could tip the world back into another deep downturn'.

So they've noticed—at least the technocrats of global governance have. The more pertinent question, however, is whether they have actually been able to do anything to head off a second and possibly even more far-reaching global economic crisis. This takes us to politics and politicians—in this case, the leaders of the countries that belong to the so-called Group of 20. The G20, as the body is usually called, is the new overarching 'steering committee' of global economic governance set up in a hurry, almost a panic, in the crisis conditions of autumn 2008 and subsequently tasked with presiding over and directing the global political economy.

The major institutions of global economic governance unquestionably knew what they were doing in issuing their worried forecasts about the future in the first half of 2016. They were seeking in good part to shape the agenda of the G20 summit that was scheduled to take place in September 2016 in Hangzhou in China. The G20 had never met before in China at the leaders level and it was hoped by many analysts and observers, including some no doubt at or near the top of the global institutions themselves, that China in its year holding the Presidency could somehow step up to the plate and restore vitality and direction to the G20's management of the global economy. After all, China represents the success of a completely different model of development from the hitherto dominant paradigm known as the 'Washington Consensus'. It has achieved a miracle of economic growth but still faces grave, and arguably worsening, environmental challenges. The Hangzhou summit was, in theory at least, the perfect opportunity to chart a way through the perfect storm.

And yet was the opportunity taken? The first two meetings of the G20 Finance Ministers and Central Bank Governors that met under the Chinese Presidency—in Shanghai in February 2016 and in Washington DC in April 2016—produced two (almost identikit) communiqués of quite astonishing complacency. Sure, the ministers and governors acknowledged the uneven character of the modest growth that was currently being achieved, but, for the rest: well, they were committed to using all available policy tools (monetary, fiscal and structural); were indeed pressing on with structural economic reforms; had not taken their eye off the Basel III banking regulations and financial sector reform; and were still trying to get countries signed up to their Base Erosion and Profit Sharing project designed to foster a fairer international tax system.

Their overall message was clear: the global economy was struggling somewhat, but, hey, we know what we're doing, and we are doing quite a lot already, thankyou very much, to manage the system. There was no hint of the possibility of a 'coming crisis' and scarcely any reference to what has, tellingly, been called the 'other crisis' (by Martin Craig in this volume). On this front, the February meeting did no more than 'welcome' the 2015 Paris Agreement on Climate Change and the April meeting merely asked the G20 Green Finance Study Group to come up with some specific options for developing green banking and integrating environmental factors into the operations of financial markets.

The next G20 Finance Ministers and Central Bank Governors meeting took place in China itself (in Chengdu in July 2016) and its communiqué did admittedly express rather more urgency about the then condition of the global economy than had previously been the case. As for the Hangzhou summit itself, the Chinese leadership was reluctant to press too hard in support of offsetting shifts of economic policy by the leading Western states. China was and is still getting used to its G20 role and, in any case, is fully implicated in the structural problems of Western economies by virtue of its banks and government agencies having bought up so much of their national debts.

China did, however, manage to secure G20 agreement in the final summit communiqué on a proclaimed new 'Hangzhou Consensus' amongst the world's leading states. In this statement, the G20's leaders committed themselves to:

> work to ensure that our economic growth serves the needs of everyone and benefits all countries and all people, in particular women, youth and disadvantaged groups, generating more quality jobs, addressing inequalities and eradicating poverty so that no one is left behind.

These were of course only words. But, they were important words because they laid down a marker at least on behalf of the cause of 'inclusive growth' (to be contrasted with the highly 'exclusive growth' typically generated in so many countries, China included, during the years of the dominance of neoliberalism).

So, what's actually going on in global governance? Why hasn't it been able to achieve more and offer bolder proposals for reform? Can it perhaps do better in the future on the back of something like the 'Hangzhou Consensus'? Or will it instead be undermined by its existing limitations and failures and soon ripped apart by the predations of 'Trumpian' economic nationalist politics in the US and elsewhere? I identify at least three important problems that limit the capacity of the current regime of global governance. They create a significant 'global governance deficit' in the very space where we presently need imagination, substance and global leadership but increasingly find instead narrow-mindedness, shallowness and an obsession with the home front.

The first problem has long been identified within the best academic accounts of the major global economic institutions. Although there has always been more internal disagreement within the major global economic institutions than some, particularly on the left, have generally wanted to admit, the fact remains that the vast majority of the staff of the IMF and World Bank especially have been trained in the US economics mainstream. Even when they dissent, as in the recent suggestion by members of the IMF research department that neoliberalism might perhaps have been 'oversold' (Ostry et al. 2016), they still tend not to stray too far from that mainstream.

What Lagarde, Mann and most global governance technocrats want is for the global economy, broadly as currently constituted, to work better than it is at the moment, to grow again and perhaps be put to the service of rather more people than was the case during the wild, expansionary years preceding the 2007–08 crisis. That is why they have been speaking out in the way they have been lately. But they don't generally want or see the need for a structurally different type of global political economy. They think it can be made to perform more effectively and more inclusively without major rebuilding.

The second problem is that the structure of global governance assembled over the years since 1944 (amidst what I have described elsewhere (Buzdugan and Payne 2016) as a 'long battle' over the nature and shape of global governance between contending groups of states) is actually very weak. The IMF and the World Bank have very little direct power over countries unless and until a country runs into economic problems and needs outside financial help. At this point, 'conditionalities' kick in and the powers of these two agencies at least are booted up. For its part, the OECD does not even have recourse to these possibilities.

Most of the time, then, all the institutions can do is seek to adjust the climate of opinion within which key global political leaders act (by, for example, issuing warnings and setting out alternate possibilities that gently critique the status quo). It is, in the end, their weakness, rather than their strength, which is their most striking feature.

By contrast, the G20 states do have the power to act. Their economies make up the bulk of the global economy. They can deploy a fuller range of economic powers, stop squeezing the life out of their economies, respond to the case for more inclusive forms of growth and begin to address the interface between renewed economic growth and the climate. It's just that they haven't as yet done these things! Admittedly, building any global governance institution is a hard task and, as I have argued previously (Payne 2014), the G20 as an organisation is deficient in design and has consequently disappointed in performance over the last few summits.

But that's not really the core explanation of the failure of world leaders to grapple imaginatively with the prospect of a further economic and financial crisis that then becomes entwined with an ecological crisis. The key point is that since its formation the G20 has always been dominated politically by a wedge of longstanding Group of 7 (G7) neoliberal states. Other countries with different political positions and traditions remain—for the moment at least—unwilling to tangle with them too openly within the structures of global governance. As a result, the G20 has stalled as a political agency capable of directing the global economic institutions. It is certainly possible that this may now begin to change because President Donald Trump is far from being a conventional neoliberal. He is, he says, a 'deal-maker'. But quite how he will interact with other G20 leaders at this year's summit, which is to be held in Germany and convened therefore by Angela Merkel, and at other summits in future years, is at this moment, and to be frank, anyone's guess.

This last observation does, however, bring us to the third and final problem that I want to mention—maybe, in fact, it's the biggest problem of all. It's also never talked about. The truth is that the leaders of the neoliberal states have not wanted and still don't want effective global governance. Why not? Because effective global governance would be public governance, would guide and regulate, would insist on controlling the wilder excesses of finance and capitalism generally, and would seek to steer the global political economy. It would necessarily be broadly social democratic in character and ethos. It could not be otherwise, given what needs to be done. However, the world's leaders over the past forty years or so have, with a few honourable exceptions, not wanted this type of global governance. They have acknowledged the need for there to exist commonly understood 'rules of the game' in finance, investment and trade, but they have much preferred the privatised, corporate, style of global governance that emerged piecemeal from the way that the original Bretton Woods institutions adapted to the end of the post-1945 consensus and the subsequent rise of neoliberalism to the hegemonic position in the global economic thinking that it enjoyed until the financial crisis.

In a nutshell, global governance can only be what the most powerful countries in the world allow it to be. The problem is that, even as presently constituted after the 'long battle', this may not be enough to avert a further global economic crisis. Actually existing global governance is simply not strong enough. One might think, therefore, that the task at hand is to consider urgently how best to strengthen it, how to 'reglobalise' as opposed to 'deglobalise' (Payne 2017). The former prospect should not be taken off the table because, until we can imagine what better global governance might look like in practice, there will be little, if any, likelihood of bringing forward the politics that might make it possible. But it's not difficult to see that the obstacles in the way of the realisation of such a vision are considerable.

Nevertheless, we know that politics changes quickly and that new leaders can bring forward new ideas. For example, China's leader Xi Jinping deliberately sought to pick up the mantle of inclusive globalisation when he spoke to the most recent gathering of the World Economic Forum in Davos in January 2017 and he gained the warmest of receptions for so doing. Yet, even as Xi was travelling to and from Switzerland, Trump was preparing to take office in the United States, committed to a strongly nationalist project of 'deglobalisation'. Somewhere between these two visions lies the immediate future of global economic governance. The 'long battle', as I have described it, would appear not to be over and, unfortunately, there is just as much a prospect of retreat as advance. We certainly can't be confident that the global governance we have will work effectively enough to head off a further global crisis.

References

Buzdugan, S., & Payne, A. (2016). *The long battle for global governance.* London: Routledge.

Ostry, J. D., Loungani, P., & Furceri, D. (2016). Neoliberalism: Oversold? *Finance & Development, 53*(2).

Payne, A. (2014, June). *The global governance of global crisis: Why the G20 was created and what we still need it to do* (SPERI Paper No. 17).

Payne, A. (2017). 'De-globalisation', or 're-globalisation'? *SPERI Comment blog.* Retrieved from: http://speri.dept.shef.ac.uk/2017/01/23/de-globalisation-or-re-globalisation/.

The Coming and Current Crisis of Indecent Work

Genevieve LeBaron

Abstract Key trends in the global labour market since the 2008 global financial crisis highlight a current and coming crisis of indecent work. Around the world insecure and precarious work is increasing, wages are stagnant and declining, and severe exploitation within global supply chains including forced labour, human trafficking, and modern slavery is growing in prevalence. Furthermore, the mechanisms to protect workers are weak and are increasingly being shaped by private actors in the global economy, and not by states. These are all signs of an escalating global labour and employment crisis. To prevent another economic crisis, the spread and normalisation of indecent work must be addressed.

Keywords Labour market · Indecent work · Employment · Global supply chains · Insecurity

G. LeBaron (✉)
SPERI and Department of Politics, University of Sheffield,
Sheffield, UK

C. Hay and T. Hunt (eds.), *The Coming Crisis*, Building
a Sustainable Political Economy: SPERI Research & Policy,
DOI 10.1007/978-3-319-63814-0_6

Global capitalism's promise was to pull people out of poverty by creating decent work. It hasn't delivered, and an escalating jobs crisis is now at the centre of the global economy's 'gathering storm'.

As of 2015, over 197 million people in the world were unemployed—which, according to the International Labour Organisation (ILO) comprised nearly '1 million more than in the previous year and over 27 million higher than pre-crisis levels' (ILO 2016: 1). The Organisation for Economic Cooperation and Development (OECD) has warned that long-term unemployment in OECD member countries has increased by 85 per cent since the global financial crisis (Inman 2014). Youth unemployment rates are especially high, hovering around 50 per cent in several countries including Spain, Greece and South Africa in 2015 (OECD 2015a). According to the ILO, 'almost 43 per cent of the global youth labour force is still either unemployed or working yet living in poverty' (ILO 2015a).

Of those who have managed to find work in the global economy, few have found jobs that are secure. More than 75 per cent of the global workforce is in temporary, short-term, informal, or unpaid work and only 25 per cent of workers are on permanent contracts (Allen 2014). These precarious and often informal jobs also tend to be lower paid; for instance, a recent OECD study documents that non-standard jobs are typically lower paid than traditional permanent work (OECD 2015b: 31). Vulnerable employment, in which workers are subject to high levels of precariousness, now comprises 46 per cent of total employment and is expected to grow by 25 million workers by 2019 (ILO 2016).

Even in the formal economy, wages across many countries and sectors have persistently declined in recent decades. Today, high proportions of those who are working are still struggling to make enough money to secure the basic necessities of life. In 2013, 837 million people—more than half of the workforce—in the G20's 'emerging countries' were working, but continued to live below or around the poverty line (Donnan and O'Connor 2014). According to the ILO in 2015, 'an estimated 327 million employed people were living in extreme poverty (those living on less than US$1.90 a day in PPP terms) and 967 million in moderate or near poverty (between US$1.90 and US$5 a day in PPP terms)' (ILO 2016: 2).

In rich countries, too, the number of people who are living in poverty, despite working—sometimes full time, or even with multiple jobs—has risen sharply. One recent study found that London's 'working poor' has increased by 70 per cent over the last decade (Hill 2015). In the US, the Obama administration's Commerce Department estimated in 2013 that total wages and salaries were 'lower than in any year previously measured' (Norris 2014).

The exacerbation of already depressed wages through 'wage theft', the illegal underpayment of workers by employers, has become widespread—even in rich countries with strong regulatory infrastructure like the UK, US and Canada (Doward 2016; Meixell and Eisenbrey 2014; Mojtehedzadeh 2016). Low-paid workers are especially vulnerable to exploitation. A study of the garment sector in Leicester found that 'standard workers earn just £3 an hour' (less than half of the UK minimum wage) (BBC 2015). Another study of low waged workers in three US cities—Chicago, Los Angeles, and New York City—found that the average worker is losing $2634 annually due to wage theft, roughly 15 per cent of their earnings (Bernhardt et al. 2009). In the US, in 2012 alone, the government and private attorneys recovered nearly US $1 billion in wages stolen by employers (Meixell and Eisenbrey 2014).

Wage theft and withholding is also an endemic feature of global supply chains. In the footwear industry alone, the practice is estimated to cost workers US $27 million per year (British Standards Institution 2016). And importantly, these illegal practices are not limited to the developing world. To name one of dozens of recent examples, the *Financial Times* recently reported that British retailer Argos was ordered in March 2017 to pay back £2.4 million to employees in back pay after they were found to have been illegally paid below the minimum wage (Vandevelde 2017).

In addition to 'regular' labour exploitation, severe forms of labour exploitation including forced labour and human trafficking are thriving. The ILO estimated in 2012 (the latest year for which an estimate is available) that 21 million people in the global economy are victims of forced labour, producing annual profits of US $150 billion for businesses (ILO 2014). These include mainstream businesses in the formal economy, such as those making and selling electronics, building new World Cup football stadiums, farming shrimp and dozens of other commodities. Although forced labour is illegal in almost every country, it seems to be becoming a stable and predictable feature of many types of global supply chains. Again, severe labour exploitation is not limited to the developing world. Forced labour has been documented within many industries in the US, Europe, Canada and other parts of the advanced capitalist world, including in agriculture, domestic work, garment production and construction (Allain et al. 2013; Phillips 2013; Crane 2013). Severe labour exploitation also affects children. In 2015, the ILO estimated that 168 million children (or 11 per cent of the world's child population) were ensnared in child labour, including large swathes who perform hazardous work or the 'worst forms' of child labour (ILO 2015b).

These are but a few signs of the escalating global employment crisis—a crisis that is increasingly acknowledged by politicians and global governance organizations, who agree that things are getting worse fast and that this poses real and long-term dangers to global economic stability.

Sparking the recent wave of cautions, the ILO warned in 2015 that 'the global employment outlook will deteriorate in the coming five years', particularly, in emerging and developing economies (ILO 2015c). A recent joint report by the ILO, OECD and World Bank warned that if current trends continue, the lack of good quality jobs in both rich and poor countries would create 'many more years ahead of weak economic growth and a 'vicious cycle' that would prove hard to break out of' (Donnan and O'Connor 2014). Even McKinsey, a global management consulting firm, has noted, 'Strains on the global labor force are becoming painfully evident. Market forces will fail to resolve demand and supply imbalances for tens of millions of skilled and unskilled workers' (Dobbs et al. 2012). It is clear that the current crisis of indecent work is escalating, and alongside the other dangers documented in this volume, could soon become much, much worse, worsening economic hardship for millions around the world and contributing to a larger and more systemic socio-economic crisis.

As such governments and international organizations are busy discussing the mounting dangers that the scarcity of decent work poses to the global economy—including downward pressure on consumer spending and growth, rising inequality, and lower living standards for the majority of the world's population. But surprisingly, little attention is being dedicated to the causes of the crisis, which needs to be faced head-on if it is to be stopped in its tracks. Three of the most important causes are worthy of particular attention.

The first is that even amidst surging corporate profits—a significant economic trend of the pre-crisis period—workers are taking home a smaller and smaller percentage of the pie. For instance, *The New York Times* reported in 2014 that 'corporate profits are at their highest in at least 85 years' at the same time as 'employee compensation is at the lowest level in 65 years' (Norris 2014). One study of value distribution along Apple's iPhone supply chain found that while Apple takes home 58.5 per cent in profits, all of the workers in its global labour force share a mere 5.3 per cent (Kraemer et al. 2011). In other words, corporations are posting record profits and are passing a tinier and tinier fraction of those profits onto the workers that make a significant contribution to their generation.

The link between soaring corporate profits and the spread and normalization of indecent work is rarely acknowledged, but lies at the heart of the crisis of indecent work. The stockpiling of mind-blowing sums of cash by global manufacturing and retail corporations' is a key cause of the mounting subsistence crisis amongst the workers in their global supply chains. In May 2016, the *Financial Times* reported that US-based companies were hoarding US $1.7 trillion in cash reserves (Platt 2016). Technology giant Apple alone, for instance, had US $203 billion in cash reserves in 2015 (La Monica 2015).

There is mounting evidence that this highly uneven distribution of value across the actors in supply chains is also fueling the demand for sub-minimum wage labour, including forced labour. In many industries, as retailers have used their market power to impose tight contracts onto and demand low-cost orders from suppliers, businesses towards the bottom end of supply chains have sought to lower costs. Amidst rising and volatile commodity prices, one of the most significant costs (which is also possible to lower, unlike oil or palm oil) is labour. In this context, businesses across many industries have sought to lower costs of labour through coercion and other illegal means. As Andrew Crane explains, 'value distribution along the supply chain, insofar as a particular stage is associated with very low-value capture, can provide significant pressure towards slavery' (2013: 54). The global jobs crisis cannot be averted without redistributing value away from corporations and into workers' pockets.

Second, the lack of government enforcement of labour standards—across both rich and poor countries—is creating a context in which businesses can enact labour exploitation with impunity. In the US, analysis of labour standards investigations by the Economic Policy Institute concluded that 'the average employer has just a 0.001 per cent chance of being investigated in a given year' (Lafer 2013). Most other countries fare little better.

Although governments frequently portray the employment crisis to be a result of abstract market forces, the reality is that it is rooted in national and global political economic policy implemented over recent decades. Most states have gutted their labour inspectorates, eliminating a crucial source of protection for workers' rights, as well as other institutions such as social welfare services that protected workers from exploitation by giving them sufficient resources to survive, even if they said 'no' to dangerous or exploitative work. As part of the broader policy shifts associated with neoliberalization (Peck 2013), governments have empowered businesses to reduce their own responsibility and liability for workers. That work has become less secure, less well paid, and more exploitative in the face of such policies is hardly surprising. Indecent work thrives where governments do not enforce the laws that make it decent—such as those protecting the payment of the minimum wage, health and safety laws, and the right to collective action.

Third, and related to the preceding point, over the last three decades governments have devolved authority and discretion over labour standards to individual companies. Companies have enacted a range of private, voluntary corporate social responsibility (CSR) programmes and initiatives including codes of conduct, ethical certifications and privately organized and managed audits of their supply chains. These programmes are now so widely accepted as legitimate means to govern working conditions that recent public legislation such as the UK's Modern Slavery Act expand and reinforce corporate social responsibility programmes by requiring companies to disclose any voluntary efforts they are making to address and prevent forced labour, human trafficking and modern slavery in their supply chains. However, in spite of the explosion of CSR programmes, there is little evidence to suggest that they actually work. Although they may do some good, only rarely do they tackle the root causes of labour exploitation in supply chains, such as low and withheld wages, the denial of freedom of association and the right to collective bargaining—in short, workers' rights. These are too often left off the table for change.

Exploitation thrives in a context where businesses set their own rules of the game and then can choose whether or not to follow them with relative impunity. In the context of the devolution of responsibility and authority for working conditions away from governments and onto individual employers, it is unsurprising that fewer employers are creating jobs that can be described as 'decent work'. History tells us that they have only tended to do so when they are made to, by governments or through pressure from workers' activism such as strikes and collective bargaining, all of which are challenging to mount and sustain in the current labour market environment.

Three crucial starting points, then to address the escalating employment crisis, are to enforce the public rules that are already on the books, to create new ones that redistribute profit from corporations to workers, and for governments to resume responsibility for labour standards, in part, by rebalancing the power relations between workers and employers. Of course, these are only starting points and they would be of most help to those already in work. We also need to confront wealth distribution where those at the very top enjoy a disproportionate share, rising inequality and unemployment in much broader terms if we are to decelerate the spread of indecent work.

After all, the problems in the global labour market and the normalization of indecent work are fundamentally political problems. They are not technical problems that can be resolved through one-off corporate social responsibility programmes or public initiatives and campaigns, or tinkering around the edges of global supply chains with social programmes. We need a system of labour and social protection that is designed to meet the challenges of today's global labour market, including complex global supply chains and the fissuring of corporate ownership and liability. Failure to address the spread of these issues and the dangers they pose to the global economy make it a certainty that if a new crisis hits indecent work will be at its heart.

References

Allain, J., Crane, A., LeBaron, G., & Behbahani, L. (2013). *Forced labour's business models and supply chains*. York: Joseph Rowntree Foundation.

Allen, K. (2014, May 19). Most of the world's workers have insecure jobs, ILO report reveals. *The Guardian*. Retrieved from https://www.theguardian.com/business/2015/may/19/most-of-the-worlds-workers-have-insecure-jobs-ilo-report-reveals.

BBC News. (2015, February 20). *Leicester textile workers paid £3 an hour, study finds*. Retrieved from http://www.bbc.co.uk/news/uk-england-leicestershire-31531924.

Bernhardt, A., Milkman, R., Theodore, N., Heckathorn, D., Auer, M., DeFilippis, J., et al. (2009). *Broken laws, unprotected workers: Violations of employment and labor laws in America's cities*. Retrieved from http://nelp.3cdn.net/e470538bfa5a7e7a46_2um6br7o3.pdf.

British Standards Institution. (2016, March 23). *BSI's Global Supply Chain Intelligence report reveals 2015 top supply chain risks.* Retrieved from https://www.bsigroup.com/en-GB/about-bsi/media-centre/press-releases/2016/March/BSIs-Global-Supply-Chain-Intelligence-report-reveals-2015-top-supply-chain-risks-/.

Crane, A. (2013). Modern slavery as a management practice: Exploring the conditions and capabilities for human exploitation. *Academy of Management Review, 38*(1), 49–69.

Dobbs, R., Madgavkar, A., Barton, D., Labaye, E., Manyika, J., Roxburgh, C., et al. (2012). *The world at work: Jobs, pay, and skills for 3.5 billion people.* McKinsey & Company. Retrieved from http://www.mckinsey.com/global-themes/employment-and-growth/the-world-at-work.

Donnan, S., & O'Connor, S. (2014, September 9). World's leading economies warned over 'global jobs crisis.' *The Financial Times.* Retrieved from https://www.ft.com/content/39c266f4-3826-11e4-a687-00144feabdc0.

Doward, J. (2016, March 12). Low-paid workers report sharp rise in 'wage theft.' *The Guardian.* Retrieved from https://www.theguardian.com/money/2016/mar/12/wage-theft-cases-low-paid-workers.

Hill, D. (2015, October 21). Number of London's 'working poor' surges 70% in 10 years. *The Guardian.* Retrieved from https://www.theguardian.com/society/2015/oct/21/number-of-londons-working-poor-surges-70-in-10-years.

Inman, P. (2014, September 3). Long-term unemployment almost double pre-financial crisis level—OECD. *The Guardian.* Retrieved from https://www.theguardian.com/business/2014/sep/03/long-term-unemployment-double-financial-crisis-level-oecd.

International Labour Organization. (2014). *Profits and poverty: The economics of forced labour.* Retrieved from http://www.ilo.org/global/about-the-ilo/newsroom/news/WCMS_243201/lang–en/index.htm.

International Labour Organization. (2015a). *Youth employment crisis easing but far from over.* Retrieved from http://www.ilo.org/global/about-the-ilo/newsroom/news/WCMS_412014/lang–en/index.htm.

International Labour Organization. (2015b). *World report on child labour 2015: Paving the way for decent work for young people.* Geneva: International Labour Office.

International Labour Organization. (2015c). *World employment and social outlook: Trends 2015.* Geneva: International Labour Office.

International Labour Organization. (2016). *World employment social outlook: Trends 2016.* Retrieved from http://www.ilo.org/wcmsp5/groups/public/—dgreports/—dcomm/—publ/documents/publication/wcms_443472.pdf.

Kraemer, K., Linden, G., & Dedrick, J. (2011). *Capturing value in global networks: Apple's iPad and iPhone.* Retrieved from http://pcic.merage.uci.edu/papers/2011/value_ipad_iphone.pdf.

La Monica, P. (2015, July 22). Apple has $203 billion in cash. Why? *CNN Money*. Retrieved from http://money.cnn.com/2015/07/22/investing/apple-stock-cash-earnings/.

Lafer, G. (2013). The legislative attack on American wages and labor standards, 2011–2012. *Economic Policy Institute*. Retrieved from http://www.epi.org/publication/attack-on-american-labor-standards/.

Meixell, B., & Eisenbrey, R. (2014). *An epidemic of wage theft is costing workers hundreds of millions of dollars a year*. Economic Policy Institute. Retrieved from http://www.epi.org/publication/epidemic-wage-theft-costing-workers-hundreds/.

Mojtehedzadeh, S. (2016, February 16). Sweet dream turns sour for victim of wage theft. *Toronto Star*. Retrieved from https://www.thestar.com/news/canada/2016/02/16/sweet-dream-turns-sour-for-victim-of-wage-theft.html.

Norris, F. (2014, April 4). Corporate profits grow and wages slide. *The New York Times*. Retrieved from https://www.nytimes.com/2014/04/05/business/economy/corporate-profits-grow-ever-larger-as-slice-of-economy-as-wages-slide.html?_r=0.

OECD. (2015a). *Youth unemployment rate (indicator)*. Retrieved from https://data.oecd.org/unemp/youth-unemployment-rate.htm.

OECD. (2015b). *In it together: Why less inequality benefits all*. Retrieved from http://www.keepeek.com/Digital-Asset-Management/oecd/employment/in-it-together-why-less-inequality-benefits-all_9789264235120-en#page3.

Peck, J. (2013). *Constructions of neoliberal reason*. Oxford: Oxford University Press.

Phillips, N. (2013). Unfree labour and adverse incorporation in the global economy: Comparative perspectives from Brazil and India. *Economy and Society*, *42*(2), 171–196.

Platt, E. (2016, May 20). US companies' cash pile hits $1.7tn. *Financial Times*. Retrieved from https://www.ft.com/content/368ef430-1e24-11e6-a7bc-ee846770ec15.

Vandevelde, M. (2017, February 16). *Argos fined for underpaying thousands of staff on minimum wage*. Retrieved from https://www.ft.com/content/51d0bc7c-f468-11e6-8758-6876151821a6.

The Coming Crisis of Planetary Instability

Peter Dauvergne

Abstract The earth is spinning into an ever-greater ecological crisis. Yet the primary solutions to end this crisis – ranging from international environmental agreements to national laws to corporate codes of conduct to individual lifestyle changes – are failing to make significant headway in ending this escalating crisis. The failure to confront rising rates of over-consumption, unequal consumption, and inequality of wealth are critical reasons for this crisis. Mainstream solutions such as the 2015 Paris Agreement on climate change and the 2016 Carbon Offsetting and Reduction Scheme for International Aviation are doing more to enrich and protect those in power than address the innate unsustainability of global wealth creation and consumption.

Keywords Ecological crisis · Climate change · Consumption · Inequality Sustainability

The earth is careering toward full-blown planetary instability. This trajectory is a symptom of the escalating socioeconomic and political crises, from rising financial instability to increasingly extreme inequality to

---

P. Dauvergne (✉)
University of British Columbia, Vancouver, BC, Canada

C. Hay and T. Hunt (eds.), *The Coming Crisis*, Building a Sustainable Political Economy: SPERI Research & Policy, DOI 10.1007/978-3-319-63814-0_7

growing insecurity for migrants and workers. At the same time, however, with each passing year, the planet's growing ecological crisis is adding to the severity of these other crises, creating a feedback loop that is accelerating the speed and intensity of each crisis.

Solving today's social, economic, and governance crises will not be possible without averting the coming crisis of planetary instability. Yet, as we see with recent international agreements to tackle global environmental problems–such as the 2015 Paris Agreement on climate change and the 2016 Carbon Offsetting and Reduction Scheme for International Aviation–those in power are trying to turn this crisis into an opportunity to stimulate economic growth, enhance corporate profits, and make money. They are hailing voluntary, bottom-up, nonbinding, and market-oriented solutions as breakthroughs in governance: as more robust and effective than imposing timelines, targets, and penalties for noncompliance. Yet such solutions are doing little to address the nature of global wealth creation, inequality, exploitation, and unequal consumption underlying global unsustainability. Instead, they are legitimizing and normalizing the coming crisis.

Already, the earth's forests, oceans, and skies are in a catastrophic state. 2016 was the hottest year ever recorded. And for the first time, average global temperatures approached 1.5°C above preindustrial times for part of the year.

This was but one sign in 2016 of the coming crisis of planetary instability. Rivers of plastic made their way into vast, swirling eddies of garbage in the Pacific and Atlantic Oceans. Mountains of electronic waste grew even higher across the world. Smog blackened the cities of India and China. Fresh water grew scarcer across Africa. And biodiversity loss intensified in Latin America and South-east Asia.

Without doubt, 2016 was a bad year for the earth. But so was 2015. That year was the warmest ever until 2016 came along. In Indonesia alone in 2015 fires to clear land for palm oil plantations scorched more than 2 million hectares of rainforests and peatlands, releasing as much greenhouse gases as the entire Brazilian economy.

Indeed, our planet is now spinning faster and faster toward an ecological crisis big enough to cause a mass extinction of species by the end of this century. *Homo sapiens* will certainly survive; if anything, the global population will likely rise by another 4 billion or so, exceeding 11 billion in total. But life for billions of people would be perilous in a world of mass extinction, as instability in the basic functions of the earth's ecology would severely disrupt–and in some cases destroy–the capacity of socioeconomic and political systems to provide food and shelter, protect human rights, and promote community wellbeing.

States have signed more than a thousand international environmental agreements to address this mounting crisis. They have created a range of international environmental organizations, such as the United Nations Environment Programme (UNEP, founded in 1972) and the Global Environment Facility (GEF, founded in 1991). Every state has also put in place environmental agencies and policies to try to lessen environmental damage. At the same time, over the past half-century increasing numbers of international and local nongovernmental organizations have been organizing street protests, lobbying governments, shaming corporations, raising public awareness of environmental problems, and partnering with business to finance conservation (Dauvergne and LeBaron 2014).

Today, even the world's multinational corporations are claiming to be pursuing 'sustainability,' touting corporate social responsibility (CSR) as a powerful way to end the crisis. Multinational mining and timber companies are claiming to be better at managing extraction sites and community security. Brand manufacturers like Apple and Nike are claiming to be improving energy efficiency, recycling rates, and waste management. Big-box retailers like Walmart and Tesco are claiming to be auditing tens of thousands of suppliers for compliance with CSR codes of conduct. Banks like HSBC are claiming to have tightened up lending rules to raise environmental standards. Over the past few decades, a plethora of market mechanisms have also been set up to help improve corporate performance and offset the damage of pollution and degradation. And there's also now an array of certification, eco-labeling, and fair-trade schemes to offer consumers a way to try to reduce the environmental and social impacts of their lifestyles (Dauvergne and Lister 2013; LeBaron et al. 2017).

Yet, despite all of this, the world is failing to make significant progress in ending this intensifying global environmental crisis. What's going on?

There are many interlocking reasons for the failure to make significant headway. For sure, the refusal of governments to redress the legacy of European imperialism and colonialism, which devastated societies and ecosystems across much of the world, is partly to blame (Crosby 2004). So too is the violence and corruption endemic in the ever-expanding capitalist world economy. The indifference, ignorance, and greed of billions of people are big reasons, too.

Three especially powerful forces, however, are combining with mounting social and financial instability and governance failures to propel us toward full-blown planetary instability: rising rates of over-consumption; rising rates of unequal consumption; and rising rates of wealth inequality. Even in the face of a rapidly growing environmental crisis, those in power are doing almost nothing to confront these forces. Instead, they are embracing consumption, big business, and billionaires as solutions, and are hailing promises to act sometime in the future as breakthroughs in global environmental governance (Dauvergne 2016).

Just look at the 2015 Paris Agreement on climate change. The agreement is 'a huge step forward in helping to secure the future of our planet', pronounced then UK Prime Minister David Cameron (BBC 2015). A 'historic agreement' and 'a tribute to strong, principled American leadership', then US President Barack Obama chimed in (Pengelly 2015). There were 'no winners or losers. Climate justice has won and we are all working towards a greener future', tweeted Indian Prime Minister Narendra Modi. The agreement is 'fair and just, comprehensive and balanced, highly ambitious, enduring and effective,' the leader of China's negotiating team told *The New York Times* (Davenport 2015).

Journalists and pundits tripped over themselves to praise negotiators. The Paris Agreement is 'the world's greatest diplomatic success', proclaimed Fiona Harvey (2015) of the *Guardian*. 'The future is bright', Lord Nicholas Stern told the *Observer*. 'If we get this right, it will be more powerful than the industrial revolution. A green race is going on' (Vidal et al. 2015).

Really?

The Paris Agreement, which came into force in late 2016 to another round of self-congratulations, does have commendable aspects (Clémençon 2016; Falkner 2016; Dimitrov 2016; Hale 2016). It aspires to limit global warming to 1.5°C, rather than 2°C. It unites the developed and developing worlds on the need for climate action, and importantly brings aboard China, India, Europe, and Japan. It lays out reporting and ratcheting-up processes to encourage states to become more ambitious in their efforts over time. And it promises by 2020 to provide poorer countries with at least US $100 billion a year in private and public funds to develop new energy technologies, reform land management, and adapt to climate change.

Yet the Paris Agreement relies on bottom-up, largely voluntary mechanisms to mitigate climate change. Reviews and reporting procedures are mandatory. But little else. The clauses for financing, emission targets, and timetables are all nonbinding under international law. There are no penalties for failing to meet commitments. Compensation and liability claims related to loss and damage are excluded. And much of the text is imprecise and open to interpretation.

Moral suasion could possibly generate reasonable rates of compliance: all international law is soft, after all. But the coming to power of US President Donald Trump in 2017–who once tweeted that 'the concept of global warming was created by and for the Chinese in order to make U.S. manufacturing non-competitive'–hardly inspires confidence in the power of naming and shaming as a way to spur compliance and action. Nor does President Trump's appointment of Scott Pruitt to head up the US Environmental Protection Agency (EPA); the former attorney general of Oklahoma has a long record of suing the EPA, backing the US fossil fuel industry, and openly questioning the very science of climate change. And nor does President Trump's announcement in June 2017 that the United States would withdraw from the Paris Agreement.

In hindsight, former UN Secretary-General Ban Ki-moon was far too optimistic when he proclaimed in October 2016: 'Strong international support for the Paris Agreement entering into force is a testament to the urgency for action, and reflects the consensus of governments that robust global cooperation, grounded in national action, is essential to meet the climate challenge' (United Nations 2016). The agreement could even turn out to be a colossal failure, doing far more to legitimize the crisis of capitalism than avert the crisis of climate change.

Government negotiators will of course be working out the specifics of the Paris Agreement for many years to come. What we do know for sure at this point is that the agreement contains the same flaw as just about every government-led environmental solution: it's designed to enable the world's wealthiest individuals and states to become even richer. The agreement relies on markets, money, business, and technology to produce solutions. Nothing in the agreement confronts the staggering inequality, greed, opulence, and exploitation underlying climate change. In the backrooms of Paris negotiators even decided to excise the clause on international air travel, an industry with carbon emissions roughly equal to those of Germany, and rising quickly.

To yet another round of self-applause, in October 2016 the United Nations' International Civil Aviation Organization (ICAO) and the international aviation industry agreed to seek 'carbon-neutral growth' after 2020 (see ICAO 2016). Britain's minister for aviation welcomed the

'Carbon Offsetting and Reduction Scheme for International Aviation' as an 'unprecedented deal' (Harrabin 2016). The president of the ICAO described the deal as 'a bold decision and an historic moment' (Milman 2016). The head of the international airline industry association was equally effusive, saying the agreement was 'at the cutting edge of efforts to combat climate change' (Owram 2016). A spokesman for Britain's Air Transport Association went even further, claiming the agreement has 'decoupled growth in aviation from growth in emissions' (Harrabin 2016).

But, wait a minute. Not all countries signed onto the scheme: notably absent were India, Russia, and Brazil. Flights to and from these countries are exempt. Moreover, the agreement does not actually require airlines to reduce greenhouse gas pollution. The airlines did agree to aspire for greater technological efficiency. But the main way they intend to achieve carbon-neutral growth after 2020 is to charge customers a small fee to offset emissions–by, say, planting trees in Indonesia or Kenya. The agreement does not even spell out clear procedures for offsetting. What is clear is that under this agreement the number of aircraft will keep rising. The number of passengers will keep rising. The consumption of jet fuel will keep rising. And the total amount of carbon emissions from international aviation will keep rising.

The Paris Agreement on climate change and the Carbon Offsetting and Reduction Scheme for International Aviation are typical of what governments now call 'sustainable development.' The underlying priority is not planetary stability, but more production efficiency, corporate profits, economic growth, and investment in technology. No one in power wants to discuss how efficiency gains can rebound into even greater environmental destruction. Most of what developed and emerging economies are calling sustainable development relies on importing unsustainable amounts of natural resources from the world's poorest and most vulnerable regions. And almost always billionaires, multinational corporations, and powerful states benefit disproportionately. Already, the extremes of inequality are obscene: the world's 8 richest men, as Oxfam International (2017) calculated, have as much wealth as the bottom half of humanity.

Of course, the Paris Agreement and the international aviation scheme are just two of a wide diversity of international environmental agreements now in place. To be fair, at least to some extent these are helping to improve global environmental management. So are government policies, corporate codes of conduct, certification initiatives, eco-markets, and eco-consumerism.

The global environmental crisis would certainly be even worse without all of these agreements and schemes. Yet, this does not change the fact that, adding everything up, the advances in environmental management are not coming close to keeping pace with–let alone reducing–the rising environmental costs of the global political economy. At the same time global environmental governance, as we see with the Paris Agreement and the international aviation scheme, has increasingly come to reflect the short-term interests and concerns of those with the greatest wealth and power–a trend with deep consequences for the nature and effectiveness of the global environmental movement more generally.

Given this, by the end of this century we'll be lucky to stop global warming at 4°C, and we could well be on our way to an earth-shattering 6°C. The rich may well think it's possible to solve every crisis by becoming even richer. Perhaps this explains the standing ovation for the Paris negotiators in 2015 and the celebratory toasts to the aviation industry in 2016. Yet there is no way around the fact that the unsustainability of global consumption, extreme inequality, and wealth creation are powerful drivers of the coming crisis of planetary instability. Ignoring this fact will surely bring ecological calamity over the course of this century. And this will certainly mean that the coming social, economic, and political crises will be even more intense, violent, and unjust.

References

BBC. (2015, December 12). COP21 Climate change deal a huge step, Says David Cameron. *BBC News*. At http://www.bbc.com/news/uk-35085810.

Clémençon, R. (2016). The two sides of the Paris climate agreement: Dismal failure or historic breakthrough? *The Journal of Environment & Development, 25*(1), 3–24.

Crosby, A. W. (2004). *Ecological imperialism: The biological expansion of Europe, 900–1900 (new edition)*. Cambridge: Cambridge University Press.

Dauvergne, P. (2016). *Environmentalism of the rich*. Cambridge, MA: MIT Press.

Dauvergne, P., & LeBaron, G. (2014). *Protest Inc.: The Corporatization of Activism*. Cambridge: Polity Press.

Dauvergne, P., & Lister, J. (2013). *Eco-business: A big-brand takeover of sustainability*. Cambridge, MA: MIT Press.

Davenport, C. (2015, December 12). Nations approve landmark climate accord in Paris. *New York Times*. At https://www.nytimes.com/2015/12/13/world/europe/climate-change-accord-paris.html?_r=1.

Dimitrov, R. S. (2016). The Paris Agreement on climate change: Behind closed doors. *Global Environmental Politics, 16*(3), 1–11.

Falkner, R. (2016). The Paris Agreement and the new logic of international climate politics. *International Affairs, 92*(5), 1107–1125.

Hale, T. (2016). 'All hands on deck': The Paris Agreement and nonstate climate action. *Global Environmental Politics, 16*(3), 12–22.

Harrabin, R. (2016, October 7). Aviation industry agrees deal to cut CO2 emissions. *BBC News.* At http://www.bbc.com/news/science-environment-37573434.

Harvey, F. (2015, December 14). Paris Climate Agreement: The world's greatest diplomatic success. At https://www.theguardian.com/environment/2015/dec/13/paris-climate-deal-cop-diplomacy-developing-united-nations.

ICAO. (2016). Carbon Offsetting and Reduction Scheme for International Aviation (State Signatories as of 12 October, 2016), United Nation's International Civil Aviation Organization. At http://www.icao.int/environmental-protection/Pages/market-based-measures.aspx.

LeBaron, G., Lister, J., & Dauvergne, P. (2017). Governing global supply chain sustainability through the ethical audit regime. *Globalizations.* At http://dx.doi.org/10.1080/14747731.2017.1304008.

Milman, O. (2016, October 6). First deal to curb aviation emissions agreed in landmark UN accord, *The Guardian.* At https://www.theguardian.com/environment/2016/oct/06/aviation-emissions-agreement-united-nations.

Owram, K. (2016, October 6). Airlines claim 'Paris moment' with global pact to limit emissions, 'historic' first for a single industry. *The Financial Post.* At http://business.financialpost.com/news/transportation/airlines-reach-global-pact-to-limit-emissions-historic-first-for-a-single-industry.

Oxfam International. (2017, January). *An economy for the 99%* (Oxfam Briefing Paper), at https://www.oxfam.org/sites/www.oxfam.org/files/file_attachments/bp-economy-for-99-percent-160117-en.pdf.

Pengelly, M. (2015, December 12). Obama praises Paris climate deal as 'tribute to American leadership'. *The Guardian.* At https://www.theguardian.com/us-news/2015/dec/12/obama-speech-paris-climate-change-talks-deal-american-leadership.

United Nations. (2016, October 5). Paris Climate Agreement to enter into force on 4 November, Department of Public Information. At http://www.un.org/sustainabledevelopment/blog/2016/10/paris-climate-agreement-to-enter-into-force-on-4-november/.

Vidal, J., Goldenberg, S., & Taylor, L. (2015, December 13). How the historic Paris deal over climate change was finally agreed. *The Observer.* At https://www.theguardian.com/environment/2015/dec/13/climate-change-deal-agreed-paris.

The European Migrant Crisis and the Future of the European Project

Nicola Phillips

Abstract The relentless news stories of recent years about the devastating numbers of lives lost in the Mediterranean, the living conditions endured by migrants seeking passage into and across Europe, and the incoherent political responses of European leaders, lead us to conclude that the European migrant crisis has been a case study in European political failure. The continued absence of effective responses to what are, clearly, inordinately complex problems reflects many of the wider political challenges facing elites and societies in Europe. But the 'migrant crisis' and its consequences are also a dimension of the 'coming crisis'. Its political, economic and social implications are deeply troubling on their own, but all the more so when viewed alongside the long-standing economic crisis in southern Europe, and the political pressures for disintegration that surround an increasingly imperilled European project.

Keywords Migrant crisis · Europe · Austerity · Disintegration
Failure

N. Phillips (✉)
King's College London, London, UK

C. Hay and T. Hunt (eds.), *The Coming Crisis*, Building
a Sustainable Political Economy: SPERI Research & Policy,
DOI 10.1007/978-3-319-63814-0_8

Digesting the relentless news stories about the devastating numbers of lives lost in the Mediterranean, the living conditions endured by the human beings seeking passage into and across Europe, and the increasingly incoherent political responses of European leaders, there can have been few other conclusions to draw over the last couple of years but that the so-called European migrant crisis has been a case study in European political failure. The continued absence of remotely effective responses to what are, clearly, inordinately complex problems reflects many of the wider political challenges facing elites and societies in Europe. But the 'migrant crisis' is also a dimension of what the editors of this book have called a 'coming crisis', inasmuch as it brings with it a wide range of political, economic and social implications that are deeply troubling on their own, but the more so when viewed alongside the long-standing economic crisis in southern Europe and the deeply contentious and unequal politics of austerity across the region.

Let us dwell for a moment on some of the dimensions of this regional political failure. It is probably not much of an exaggeration to say that the migration crisis represents one of the most notable and consequential episodes of political failure in the history of European cooperation, which many worry retains the capacity to challenge the core of the European project. The early attempt by the German government to lead (by example) a humane regional response to the crisis by welcoming large numbers of refugees was perhaps inevitably doomed to failure, given the political conditions that are attached to issues of immigration across Europe.

In Germany itself, the so-called 'open-door' policy was short-lived, beset quickly by political tensions within Chancellor Merkel's own ranks as the logistical, bureaucratic and political realities of the policy began to bite, eventually forcing a policy retreat and a return to an idea of managed asylum. Under its replacement—a new, hard-line policy designed to neutralise political opposition from anti-immigration groups and the right wing—compulsory removals have increased, and a system of cash inducements to those prepared to leave voluntarily has been initiated (Chazan 2017). Political debate continues fiercely and acrimoniously, particularly around claims that the fabric of German society is increasingly threatened by the twin increases in the incidence of violence and hate crime.

A part of the rising political pressure within Germany resulted from Chancellor Merkel's inability to secure consensus among Germany's

European partner governments as the migrant crisis first became apparent in summer 2015. Some of the largest countries, including the UK, were unwilling to offer asylum to a proportionate share of refugees or to think in collective regional terms about approaches to addressing either the political or the logistical challenges of responding to the growing crisis. Countries such as Hungary and Macedonia could not be prevented from violently blocking the passage of migrants, who remained camped in appalling conditions at Europe's borders.

The result was a deal with Greece, struck in early 2016 as the crisis reached desperate proportions, for the deportation of refugees to Turkey in the hope of achieving a more 'orderly' admission of manageable numbers of people into European territory. Quite apart from the questions that were raised about the legality of this scheme (Gayle 2016), it was clear before its implementation that enforcing this deal was the tallest of orders. We know now from experience since implementation started that it has done little to solve the problem. The most troubling manifestation of this failure of policy is that thousands of people are known to be 'missing', especially as temporary camps have been forcibly closed in Greece, and their inhabitants thought to have been dispersed across Europe by smugglers or to be living rough away from 'official' refugee camps.

The dimensions of political failure are further evident in the implementation of those (limited) agreements which *have* been reached between European member states in relation to responsibility for receiving refugees and migrants, particularly in the lack of support given by wealthier nations to schemes aimed at mitigating its causes at source. The United Nations High Commissioner for Refugees reported in March 2016 that, while donor countries at a meeting held in London had pledged US$6bn for humanitarian and development programmes in Syria and neighbouring countries, only about 8 pe rcent of those pledges had been disbursed, and donors were still yet to allocate some of the funds announced for individual agencies (Wintour 2016). We remain in a situation, where the richest countries in the world have pledged to resettle only about 0.5 per cent of refugees from Syria. Under the programme of EU quotas for resettlement, 106,000 refugees were intended to be relocated from camps in Italy and Greece, but, as of March 2017, only 13,500 had been accepted by other European countries—a situation not inaccurately described as a 'humiliation' for the President of the European Commission who had led the development of the quotas programme (Waterfield 2017). Countries like Hungary, the Czech Republic and Poland have either refused or failed to comply with the terms of the EU agreement.

Meanwhile, the logistical crisis grows: given that some 57 per cent of applications for refugee status since mid-2015 have been or are in the process of being declined, EU countries are faced with the challenges of collectively deporting some 1.5 million people. It is estimated that only around a third will end up being deported, with the remaining one million existing on the margins of society as part of an exploding population of unauthorised migrants (Waterfield 2017).

So, how should we understand the consequences of this story of political failure, and particularly their connections with a scenario of a 'coming crisis'? The first set of consequences is, of course, for the individual human beings caught up in a situation of political limbo. It is clear that the crisis is driven in the first place by the conditions in countries like Syria which are by any human standards intolerable. The ravages of civil war, political violence and humanitarian crisis are well documented. It is known that many of the migrants who attempt the crossing to Italy are fleeing myriad forms of violence and deprivation in Libya. Many have originally migrated there from north and sub-Saharan Africa or south Asia, and have been further compelled to flee by the conditions encountered in detention camps at their destination. The conditions they endure in camps across destination points in Europe are also known to be appalling—in some instances 'unfit for animals' (Townsend 2016)—and the forcible demolitions of camps in countries such as France and Greece during 2016 and 2017 have compounded the conditions of destitution in which migrants and refugees are living.

We know who the major beneficiaries of this crisis are: the smuggling and trafficking networks which are profiting enormously from the desperation of refugees and migrants, and 'employers' who see business opportunities in the huge numbers of people denied access to labour markets in European countries. Evidence is accumulating thick and fast of the patterns of severe labour exploitation, including forced labour and child labour, which are connected with the migration crisis. In countries such as Libya, it is documented that many detention centres for north African migrants operate as 'slave markets' or 'holding pens', run by militias working with traffickers (Kington 2017). Trafficking for labour exploitation and sexual exploitation is increasingly well documented in such countries as Turkey and Lebanon, and across Europe (Business and

Human Rights Resource Centre 2016; Freedom Fund 2016). Children are known to be a 'preferred target' of criminal gangs looking to force migrants into slavery (Rankin 2016).

Second, the political consequences of the failure of political leadership are increasingly evident. The crisis has caused remarkable and novel forms of mobilisation across the region demanding of our political leaders humane generosity in their treatment of refugees fleeing war, conflict and poverty. Across Europe street demonstrations and protests, 'refugees welcome' campaigns on social media, and the publicised kindnesses of people and communities made for rare viewing in our contemporary age as the crisis first unfolded in 2015, and it is hard to call to mind any comparable recent instances of mobilisation of this sort and on this scale around the issue of immigration.

But it has also, predictably, hardened political positions across Europe at other points on the spectrum, notably among far-right groups such as Pegida, and populist groups and political parties wedded to anti-immigration and nationalist agendas. Given existing concerns about the renewed vigour of right-wing and far-right politics in the context of economic crisis, this fuelling of political tension across the region is dangerous for the coherence of the European project. This is especially so when set alongside such developments as the UK's referendum on continued EU membership, whose outcome is recognised to have pivoted, for sections of society, around perceptions of immigration more than any other issue.

Third, the economic and social consequences of this story of political failure are marked. Nowhere is this more apparent than in Greece, a country already mired in economic crisis and an extremely fragile political landscape, where the burdens of implementing the 'deal' struck by European leaders have been economically heavy. At the same time, the crisis has implied a flourishing of the informal, illegal and illicit economies, with consequences for the project of economic recovery in some of the countries still experiencing deep recession or austerity.

The social consequences of these twin economic and political crises are in this sense manifold. All of the dynamics noted above crystallise into a landscape of new and deepened inequalities—between countries; in economic, political, social terms; and relating to opportunity and rights—which must be of concern for any progressive version of the European project. If the conjunction of crisis points is indeed pointing toward the potential, at least, for accelerated European disintegration, and given the sheer scale of the political salience of migration and immigration issues across member states, then the task of finding politically sustainable responses to the crisis is indisputably a matter of not only political, but also economic, urgency.

The question, in closing, is therefore what can now be done. If the argument is accepted that what we have seen is a story of far-reaching political failure in Europe, which has given a turbo boost to populist, anti-immigration and far-right politics, then it is hard to escape the conclusion that political consensus is as far out of reach as it ever was in this context. Progressive political leadership on the migration crisis has become suffocated by these politics, with the German leadership forced into a climbdown, and the European Commission humiliated by its incapacity to achieve enforcement of its quotas programme across the region, even among the leading countries. The UK government's initiation of the process for 'Brexit' from the EU has not only pushed the UK even further out of the picture in addressing the migration crisis and the humanitarian disaster it represents, but has also provided the biggest imaginable distraction for the rest of the leading countries of the EU. Their attention is compelled to be drawn towards the monumental challenge of managing the UK's exit while simultaneously seeking to hold the European project together.

In this intra-European context, and given the barriers to striking a deal with Libya or a further deal with Turkey, attention has turned to other key transit countries. In March 2017, Chancellor Merkel's trip to Egypt and Tunisia aimed to pursue the new mantra of 'control' by 'off-shoring' the processing of asylum requests to these countries as a deterrent to migrants from attempting the crossing to Europe. But this is equally perilous terrain in political terms, and the proposals have generally not been well received in these countries.

In short, it is hard at present to see from where a political solution might be forthcoming in the near future. Yet, what remains painfully clear is that the ongoing failure to secure one can only deepen the humanitarian crisis, further imperil the European political–economic project, and heighten the sense of a coming crisis that surrounds it.

References

Business and Human Rights Resource Centre. (2016). Syrian refugees: Abuse and exploitation in Turkish garment factories, April. https://business-humanrights.org/en/modern-slavery/syrian-refugees-abuse-exploitation-in-turkish-garment-factories?dateorder=datedesc&page=0&componenttype=all.

Chazan, G. (2017, March 30). Syrian refugees in Germany: Paths diverging. *Financial Times*. Retrieved from https://www.ft.com/content/304cebc0-08c7-11e7-ac5a-903b21361b43m.

Freedom Fund. (2016). Struggling to survive: Slavery and exploitation of Syrian refugees in Lebanon, April. http://freedomfund.org/our-reports/%EF%BF%BCstruggling-survive-slavery-exploitation-syrian-refugees-lebanon/.

Gayle, D. (2016, April 2). EU-Turkey refugee plan could be illegal, says UN official. *The Guardian*. Retrieved from https://www.theguardian.com/world/2016/apr/02/eu-turkey-refugee-plan-could-be-illegal-says-un-official.

Kington, T. (2017, March 3). Terror in Libya drives new wave of med migration. *The Times*. Retrieved from https://www.thetimes.co.uk/article/terror-in-libya-drives-new-wave-of-med-migration-jjhrd5gpf.

Rankin, J. (2016, May 19). Human traffickers 'using migration crisis' to force more people into slavery. *The Guardian*. Retrieved from https://www.theguardian.com/world/2016/may/19/human-traffickers-using-migration-crisis-to-force-more-people-into-slavery.

Townsend, M. (2016, May 28). Protests grow as Greece moves refugees to warehouses 'not fit for animals'. *The Guardian*. Retrieved from https://www.theguardian.com/world/2016/may/28/greece-refugee-warehouses-not-fit-for-animals.

Waterfield, B. (2017, March 3). Step up deportations, Juncker urges as 1m flout rules. *The Times*. Retrieved from https://www.thetimes.co.uk/article/step-up-deportations-juncker-urges-as-1m-flout-rules-xhj2jtfjc.

Wintour, P. (2016, March 30). Half of $12bn refugee fund pledged at London meeting not disbursed. *The Guardian*. Retrieved from https://www.theguardian.com/world/2016/mar/30/refugee-crisis-half-funds-pledged-london-conference-david-cameron-not-disbursed.

The Paradox of Monetary Credibility

Jacqueline Best

Abstract One of the hallmarks of recent political events has been a growing suspicion of so-called elites and experts—chief among them, those very central bankers who were only recently being celebrated for saving the global economy. Central bank independence and rule-based monetary policy became the norm over recent decades on the assumption that democratic influence would erode the policies' credibility and therefore effectiveness. Yet, recent electoral events remind us that insulating economic decision-makers too much from popular concerns tends to erode their legitimacy—and thus undermines the credibility they seek to protect. Whereas central banks provided some of the most effective responses to the last crisis, it is unlikely they will have the legitimacy and effectiveness needed to fight the next crisis. In fact, their declining legitimacy may be one of its major causes.

Keywords Monetary policy · Central banks · Credibility · Rules
Legitimacy

J. Best (✉)
University of Ottawa, Ottawa, Canada

C. Hay and T. Hunt (eds.), *The Coming Crisis*, Building
a Sustainable Political Economy: SPERI Research & Policy,
DOI 10.1007/978-3-319-63814-0_9

The Brexit vote, the election of Donald Trump, and the rise of the extreme right in Europe all remind us that no matter how hard economic policymakers try to insulate their decisions from politics, they will never succeed. In fact, recent events demonstrate that not only are economic policies inherently political, but the very attempt to separate them from political pressures can very easily have the opposite effect.

One of the hallmarks of the recent sharp veer to the right has been a growing suspicion of so-called elites and experts—chief among them, those very central bankers who were only recently being celebrated for saving the global economy (Mallaby 2016). If this suspicion of expertise continues to grow, it has the potential not only to precipitate the coming economic crisis by eroding monetary credibility, but also to frustrate policymakers' attempts to respond effectively.

There are of course a great many factors driving the rise of the right, and to suggest that they can be reduced to the effects of failed economic policies would be to oversimplify in the extreme. Certain economic policies do nonetheless bear some responsibility for our present predicament—notably the last two decades' trend toward a particular kind of politically insulated, rule-based form of monetary policymaking. Central bank independence and rule-based monetary policy became the norm over the past few decades on the assumption that requiring policymakers to follow simple rules, like meeting an inflation target, would make their policies credible.

Yet, policymakers forgot that credibility is not a natural phenomenon, granted by the laws of orthodox economics, but a profoundly social and political one (Braun 2017). For a policy to be credible, it must be believed. In the context of a democratic society, that belief also depends ultimately on the legitimacy of those policies and the institutions and individuals who produce them. Recent electoral events have reminded us once again that insulating economic decision-makers too much from popular concerns tends to erode their legitimacy—and thus undermines the credibility that they seek so jealously to protect.

In order to understand how we got into this present dilemma, and to ascertain the risks of a future crisis, we need to look at three key moments in the recent history of monetary policymaking: the rise of the rule, the proliferation of exceptions, and the erosion of legitimacy.

One Rule to Bind Them All

Rule-based monetary policy and central bank independence have not always held the kind of near-divine authority that they do today. After the failure of the gold standard and its disruptive influence in the lead up to the Second World War, the task of managing the value of money was seen as a central responsibility of elected governments, to be pursued with an ever-changing mix of fiscal, wage and price control and monetary policies.

Yet, by the mid-1970s, as economic policy fell into a stop-go cycle of stimulus and restraint and inflation continued to surge, monetarism, one of the first and most influential economic theories to advocate a strict rule-based approach to managing inflation, began to gain influence.

The logic of rule-governed monetary policy was straightforward: policymakers would identify and publish their commitment to a particular monetary rule or target (today, the golden rule seems to be a 2 per cent inflation target). Market actors and the general public would then adapt their actions according to this target. Central bank independence was seen as crucial because it was believed that technocrats were more likely to stick to the rule than elected officials who might be too influenced by popular concerns (since monetary policy always produces winners and losers) (Kydland and Prescott 1977).

After the inflationary instability of the 1960s and 1970s, this rule-governed approach to policy was designed to be both politically and economically stabilizing: to do away with the problem of political uncertainty by removing not only governments' but even central bankers' discretion: just stick to the rule, and everything will work out. A tidy, efficient, depoliticized (although certainly not apolitical) approach to monetary policy.

In practice, of course, the initial effectiveness of the monetarist turn was far from universal: although the painfully high interest rates imposed by the US Federal Reserve Chairman, Paul Volcker, in the early 1980s were generally effective in bringing down inflation, in the UK, Thatcher's governments ultimately gave up on their experiment with a monetary rule and used deflationary fiscal policy and the resulting mass unemployment to bring down the inflation level (Elgie and Thompson 1998).

Over time, however, the rule-based approach gained traction on a global level, as market actors and politicians came to understand and expect that monetary rules would maintain price stability. Mainstream economists came to love rule-based monetary policy as did politicians. In the 1980s and 1990s, most central banks moved toward an increasingly rule-based approach to monetary policy, with inflation targeting becoming the norm in many countries in recent years.

Although even former US Federal Reserve Chairman, Alan Greenspan, has admitted that the remarkable price stability of the 1980s and 1990s cannot be fully attributed to the effectiveness of rule-based monetary policy, the victory of simple rules became an extremely powerful narrative (Greenspan 2004).

The Exceptions Start Piling up

Since the 2008 financial crisis, however, those rules came under increasing strain as central bankers experimented with a whole range of unconventional monetary policies in their efforts to respond.

There has been more attention of late to central bankers' increasing power and influence on the global stage, as they were given primary responsibility for responding to the crisis. However, there has been less attention to a key paradox underlying central bankers' new roles on the world stage: they are being forced to govern through exceptions in an era in which rule-following has become the ultimate source of policy credibility. Where central bankers are supposed to stick to the rules, they have found themselves endlessly making exceptions, promising that one day things will return to normal.

Governing through exceptional policies is always a politically fraught undertaking, particularly over the long term, but it is even more difficult in a context in which the dominant convention is one of strict rule-following.

Today, we are faced with a situation in which the rules no longer apply but are still being invoked as if they did. A recent Buttonwood column notes that the Bank of England has missed its inflation target 'almost exactly half the time' since 2008 (Buttonwood 2016). The European Central Bank (ECB) has effectively expanded its narrow mandate, which formally requires it to make price stability its top priority, by arguing that employment and other issues are crucial to achieving it. Yet the ECB and the Bank of England continue to act as if the old rules still apply.

If we look beyond the narrow rules that are supposed to be governing central bank actions and examine the wider changes in their recent policies, we find similar patterns. Scratch an unconventional monetary policy and you will find a kind of economic exceptionalism: an argument that the instability that we continue to face is extreme enough that it requires a radical but temporary suspension of economic rules and norms.

Most of the unconventional monetary policies that have been tried to date, and just about all of those that have been proposed as future possibilities if we face a renewed global recession, break quite radically with existing norms. Negative interest rates weren't even supposed to be economically possible (until they were tried), while quantitative easing (a central bank's buying of bonds by massively increasing the size of its balance sheet) still carries a whiff of irresponsibility linked to its past as a way for governments to avoid fiscal retrenchment by 'printing money.'

More recent proposals include 'helicoptering' money into the government's or the public's accounts, abolishing cash to make low interest rates effective, and even introducing a reverse incomes policy—a government-enforced increase in wages (as opposed to the wage controls of the 1970s) to try to get inflation going.

All of these existing and potential policies break with current economic norms, and all are being pitched as temporary, exceptional measures that are (or may be) necessary in the face of an extreme future crisis.

Ironically, rule-following was designed precisely to avoid any reliance on ad hoc monetary policies of the kind that had proliferated in the 1960s and 1970s. Yet rules only seem great until they don't apply anymore. A rule that pretends it can always apply inevitably runs into serious problems when an exception becomes necessary.

The Erosion of Legitimacy

As the exceptions have started to pile up, market actors, politicians, and the general public have begun to ask questions about the legitimacy of central bankers' considerable powers (Fleming 2015; Buttonwood 2015).

Yet, to fully understand how the legitimacy of central banks became contested, we need to go back further, to the height of the rule-based order. It is true that central bankers like Greenspan were invested with almost mythic authority during the Great Moderation (an era of unusual macroeconomic stability dating roughly from the mid-1980s to the onset of the 2008 financial crisis); yet, the low-inflation order that they sought to maintain through their monetary rules ultimately benefited some far more than others.

Central bankers made an explicit decision to focus narrowly on price stability at the expense of other potential objectives, including fostering employment and encouraging growth. Although the goal of this narrow approach was to get politics out of the mix by aiming for less contestable policy outcomes, this rule-based strategy nonetheless had political consequences.

While the rule-based approach assumes that low inflation will benefit everyone, the evidence on the issue generally points in a different direction, suggesting that growth and equality are best served with moderate rather than very low levels of inflation (Kirshner 2000; Monin 2014; Bulir 2001). There will also be winners and losers for any given interest-rate level: for example, young families with big mortgages will generally benefit from low-interest rates, while seniors living on their savings will be penalized by them. There is simply no way to avoid the distributive consequences of monetary policy.

At the same time, the very fact of excluding broader economic targets from their calculations made monetary policy blind to some of the signs of economic instability and limited their ability to predict and prepare for the 2007–2008 crisis. As late as August of 2007, for example, the Federal Reserve was still more preoccupied with the possibility of inflation than with the looming threat of financial instability (Telegraph 2013).

Of course, once the crisis hit, central bankers underwent a rapid about-face, shifting from a narrow reliance on monetary rules to an unprecedented experiment with exceptional measures. Yet, even as monetary policy shifted from rule to exception, the same dynamic was at work, as policies designed to avoid political conflicts ultimately end up exacerbating them.

Why did central banks play such a massive and extended role in responding to the financial crisis? In large measure, because elected leaders wanted them to play that role. Particularly as the immediate crisis receded, and politicians decided to switch gears from fiscal stimulus to austerity, they came to rely increasingly on central banks' exceptionalist measures (like very low interest rates and quantitative easing) to keep the economy moving. By refusing to take political responsibility for the need for further stimulus, they passed the buck to central banks.

Yet, once again, by trying to avoid political conflict, these governments ultimately created a different kind of political problem: granting central banks more authority than they could legitimately sustain over the longer term.

As central banks' legitimacy has come into question, so has their policies' long-term credibility, putting them in a very difficult place indeed. After all, how does monetary credibility work? It depends, at the end of the day, on faith. Even orthodox economists agree on this: the credibility of a central bank's monetary rules hinges on public expectations that a policy commitment will be followed through in practice—that the rule will be followed.

As the exceptions pile up, it is becoming increasingly difficult for central bankers to convince the public of the credibility of their rules. If employers, investors, and wage earners don't believe that policymakers can or will stick to the rule, then they will begin to change their monetary practices, eroding the virtuous circle of monetary credibility as, for example, wage earners start asking for higher future increases to hedge against future inflation. While the habitual nature of low inflation expectations, and the practices that they inform, will continue for some time, if central banks' legitimacy continues to erode, there is a growing risk that they will come unstuck in the context of a future shock.

Whereas central banks provided some of the most effective responses to the last crisis, it is therefore unlikely that they will have the legitimacy and the effectiveness needed to fight the next crisis. In a remarkably short period, central bankers have gone from the heroes to the villains in many political narratives. The Governor of the Bank of England, Mark Carney, has faced unprecedented attacks for the Bank's grim predictions of the likely economic fallout of a positive Brexit vote, while Congressional Republicans' post-crisis calls to 'Audit the Fed' in the USA have now been joined by President Trump's repeated criticisms of the Federal Reserve Board. It will be far more difficult for politicians to call on those same monetary institutions to help fight the next economic crisis if their credibility is in tatters.

Where should we go from here? In the short term, with authoritarian exceptionalism on the rise in the USA and elsewhere, we may be glad that there are at least a few institutions, like central banks, that are at least somewhat insulated from direct political control. Yet, the lessons of the past decade should make us think rather carefully over the longer term, as we try to understand how things could have gone so very wrong, and how to prevent them happening in the future.

The lesson to be drawn here is a simple one: while too much politics may make for poor monetary policy, too little politics can be just as dangerous over the longer term. When monetary institutions seek to insulate themselves from political pressures and concerns in order to ensure their policies' credibility, they may actually become vulnerable to more dramatic erosions to their legitimacy—and thus to their credibility.

And as recent events have reminded us once again, when our institutions lose their legitimacy, bad things often start to happen.

References

Braun, B. (2017). Speaking to the people? Money, trust, and central bank legitimacy in the age of quantitative easing. *Review of International Political Economy, 23*, 1064–1092.

Bulir, A. (2001). Income inequality: Does inflation matter? *IMF Staff Paper, 48*.

Buttonwood. (2015). *Monetary policy, politics and the economy: Central banks in the firing line*. London: The Economist.

Buttonwood. (2016). *The fallibility of central banks*. London: The Economist.

Elgie, R., & Thompson, H. (1998). *The politics of central banks*. New York: Routledge.

Fleming, S. (2015, February 24). Fed paves way to raise rates this year as US economy strengthens. *Financial Times*.

Greenspan, A. (2004). *Risk and uncertainty in monetary policy*. Washington, DC: US Federal Reserve Board.

Kirshner, J. (2000). The study of money. *World Politics, 52*, 407–436.

Kydland, F., & Prescott, E. (1977). Rules rather than discretion: The inconsistency of optimal plans. *Journal of Political Economy, 85*, 473–491.

Mallaby, S. (2016, October 20). The cult of the expert—And how it collapsed. *The Guardian*.

Monin, P. (2014). *Inflation and income inequality in developed economies* (Council on Economic Policies Working Paper).

Telegraph. (2013, January 18). Federal reserve missed financial crisis warning signs in 2007, documents show. *The Telegraph*.

Enduring Imbalances in the Eurozone

Scott Lavery

Abstract The eurozone continues to be afflicted by a number of profound imbalances. Enduring weaknesses within the eurozone's southern 'periphery', the euro's deflationary bias and entrenched patterns of uneven development together threaten to undermine the eurozone as an economic and political unit. Technical 'fixes' exist to the eurozone's imbalances. For example, enhanced fiscal transfer between member states could alleviate some of the dysfunctionalities of the single currency in its current form. But intractable political obstacles render these solutions untenable. As a result, eurozone states remain entrapped in a dysfunctional and unstable currency arrangement. When the 'coming crisis' does emerge, the imbalanced eurozone is likely to be front and centre of the global maelstrom.

Keywords Imbalances · Eurozone · Single currency · Disequilibrium Weaknesses

Jean Monnet–one of the architects of the European integration process–opined in his memoirs that 'Europe will be forged in crises, and will be the sum of the solutions adopted for those crises'. Since 2010,

S. Lavery (✉)
SPERI, University of Sheffield, Sheffield, UK

C. Hay and T. Hunt (eds.), *The Coming Crisis*, Building a Sustainable Political Economy: SPERI Research & Policy,
DOI 10.1007/978-3-319-63814-0_10

the eurozone has been mired in a protracted economic and social crisis. A nascent recovery has taken hold but no convincing 'solution' to the euro's difficulties is in sight. Whilst the policy fixes advanced by European policymakers have thus far averted a break-up of the single currency, they have also consolidated a series of structural imbalances and profound weaknesses within the emergent framework of European capitalism. In considering the risks of a coming crisis, three areas of major imbalance standout in particular.

First, the programme of fiscal austerity imposed by the Troika (the International Monetary Fund (IMF), European Central Bank (ECB) and the European Commission) has produced huge economic and social upheaval. It has been calculated that once fiscal multipliers–the 'knock-on effects' of large spending cuts–are taken into account, fiscal consolidation reduced eurozone GDP by 7.7 per cent in 2013 alone (Gechert et al. 2015: 6). The effect of this on those countries which were subject to the Troika's conditionality programmes in return for bailouts has been devastating. Between 2010 and 2016, GDP fell in Greece by 27 per cent as public and private investment dried-up. As growth plummeted, Greece's debt-to-GDP ratio rose substantially, even as it ran primary budget surpluses (OECD 2014). In 2016, Greece slipped back into recession, weighed down by an unrepayable debt burden and unable to stimulate its economy through expansionary measures.

Elsewhere on the eurozone's periphery, trouble bubbles under the surface. Portugal–often held-up as the 'good pupil' of eurozone austerity–has a government debt level of over 129 per cent. A recent IMF report emphasised that a small change in market sentiment could 'render Portugal's capacity to repay [its debts] more vulnerable' (Khan 2015). In this context, Portuguese bond yields rose to a two year high in 2016 as investors questioned the durability of the country's putative economic recovery. As a result, although growth has returned to some quarters of the eurozone, it remains in the words of the ECB president 'weak, fragile and uneven' (Draghi 2014). In a context of growing global uncertainty, this renders the eurozone vulnerable to a future downturn in market sentiment.

Second, the threat of deflation–low and falling prices–continues to haunt the eurozone's nascent economic recovery. As Jeremy Green outlines in his chapter, there is evidence that a second 'stagflationary regime' is emerging across the post-crisis advanced capitalist world. The deflationary menace has been an omnipresent feature of the eurozone crisis.

Over the past four years, inflation has consistently undershot the ECB's target of 2 per cent. At the start of 2016, headline inflation in the eurozone stood at minus 0.2 per cent. Although recent increases in energy costs have created an upwards pressure on prices, 'core' inflation—which strips out volatile changes to food and energy prices—remains consistently low at 0.9 per cent.

The ongoing threat of deflation represents a major concern for two reasons. First, consistently falling prices can lead households and firms to hold-back on their spending and investment as they anticipate future price drops. In turn, this can lead to lower growth, lower investment and further price cuts: a 'deflationary spiral', as was experienced in Japan throughout the 1990s. Second, falling prices increase the value of debt in real terms. This can act as a further disincentive to investment and therefore as a further drag on growth. Unorthodox monetary policy–sustained low and even negative interest rates and quantitative easing (QE)–has been deployed by the European monetary authorities to counteract this deflation threat. However, these instruments were deployed late (the ECB only began QE in 2015, some 7 and 8 years after the US and UK, respectively) and–given the consistent failure of the ECB to hit its inflation target–have proven inadequate. In the absence of expansionary fiscal policy, the threat of an ongoing low growth, low investment equilibrium looks set to continue, with deflation representing an ongoing threat to the stability of the eurozone's economic recovery.

Third, uneven development has deepened between the 'core' and 'periphery' of the eurozone (Streeck 2015). Amidst years of falling investment, industrial production has collapsed in the southern European periphery: output has fallen by 25 per cent in Italy, Spain and Greece compared to 2008. This has resulted in a further concentration of industrial activity in Germany (and countries integrated into its supply chain), with exports as a share of GDP rising to 50 per cent in 2013 (Becker and Jäger 2011). German exporters continue to benefit hugely from a real exchange rate which is lower than would have been the case under the Deutschmark. German firms have also drawn-in huge volumes of capital from investors fleeing turbulence within the eurozone periphery (Offe 2015).

These enduring imbalances suggest that the euro has failed to deliver upon the ambitions of its founders. Advocates of European Monetary Union (EMU) initially argued that the single currency would help to facilitate a positive convergence in the productive capacity and living standards of member states. Instead, the euro has precipitated the precise opposite, deepening polarisation between the eurozone's northern 'core' and its southern 'periphery'. Price stability, it was claimed, would support and not undermine the 'social models' of member states. However, the 'structural reforms' required by the Troika's conditionality programmes have led to swingeing public expenditure cuts and the erosion of employment protections across member states, most notably in Spain, Portugal and Greece (Heyes and Hastings 2015). In the absence of monetary sovereignty, eurozone states have been driven to secure economic recovery through 'internal devaluation', that is through cutting wages and social entitlements in attempts to induce investment and stimulate export-led growth. However, in a context of weak effective demand and anaemic investment, internal devaluation has undermined economic activity, eroding tax takes and thereby rendering deficit reduction more difficult to achieve (Stockhammer 2016).

The intractable dysfunctionality of the euro area has not led to a fundamental rethink amongst European policymaking elites of how EMU should function. Instead, the Commission has attempted to deepen the prevailing logic of the single currency through a series of reforms designed to further insulate the euro from democratic politics. For example, the 'Six-Pack' of regulations—five regulations and one directive—which were adopted in 2011 aimed to enhance the Commission's oversight of member states' domestic economic policies (Verdun 2015). The Six-Pack seeks to ensure that member states implement Commission recommendations—potentially affecting their macroeconomic positions on domestic labour law, wage policies, and social services—on pain of severe sanctions (Scharpf 2015). These sanctions are automatically implemented unless a qualified majority on the European Council intervenes.

Supranational oversight over deficit states' balance sheets has therefore been strengthened, locking-in the deflationary bias of the single currency whilst further marginalising the capacity of national politicians to secure adjustment through counter-cyclical measures. The European Semester process, which complements the 'Six-Pack', requires that member state governments pass their budgets to the Commission for review before they are communicated to their national parliaments (Scharpf 2015: 8). Entrenched and inflexible fiscal discipline, therefore, remains the enduring *modus operandi* of the euro, but this is now combined with a growing resort to 'extra-legal' forms of supranational oversight and executive rule (Oberndorfer 2015: 188). The political sustainability of this 'new economic governance' architecture—not to mention its democratic credentials—is questionable.

There is no shortage of technical fixes to the eurozone's current impasse. A plethora of alternative economic programmes has been advocated by progressive forces within the EU which challenge the disciplinary and deflationary policies favoured by incumbent EU policy-making elites. Persistently low German unit labour costs—which were stagnant between 2000 and 2007 and which were a key driver of some of the internal trade imbalances within Europe in the pre-crisis period—continue to be restrained (Flassbeck and Lapavitsas 2016). By reflating domestic demand through allowing wages to rise and allowing some inflation into the German economy, some of these internal imbalances could be corrected. However, the political will to drive through these reforms has been notably absent. German exporters and trade union confederations remain resolutely opposed to sacrificing their comparative export advantage, meaning that powerful social forces in Germany–haunted by the experience of high unemployment in the 1990s–are unwilling to 'rebalance' the eurozone through domestic reflation.

Supranational policy 'fixes', supported by political theorists such as Jurgen Habermas and by some European social democrats, Greens and European trade unions have also been advanced (Habermas 2014). For example, the EuroMemo Group has advanced a series of policy proposals, including plans to launch a European-wide green industrial strategy and to create a federal-level fiscal policy in order to facilitate redistributive transfers between member states (EuroMemo 2015). These policies could go some way towards resolving the current impasse in the eurozone insofar as they would substitute 'internal devaluation' with the possibility of discretionary counter-cyclical 'adjustment' within laggard regions. However, northern European politicians–with one eye on their domestic electorates–refuse to countenance the possibility of enhanced fiscal redistribution across European borders.

This problem is compounded by protracted resistance to increases in the EU budget across member states. It has been calculated that the EU budget would have to increase by 300 per cent if supranational fiscal transfers were to become an effective macroeconomic policy tool. However, this would translate into member states granting on average a further 7 per cent of their public expenditure to Brussels annually (in Germany the figure is 15 per cent) (Streeck 2015). With public finances already highly constrained and with palpable opposition to the prospect of 'more Europe', large redistributions of resources from national governments to the EU institutions looks highly unlikely, if not outright impossible, under present conditions.

The deflationary bias of EMU, the evident dysfunctionality of the euro and the political limits to federalist solutions has led some thinkers to countenance the abolition of the single currency altogether (Elliot and Atkinson 2016; Streeck 2014). On this reading, member states' sacrifice of monetary sovereignty in the 1990s was a fatal mistake. It removed one of the core instruments available to democratic governments: the capacity to shape their monetary policy in line with domestic conditions and to secure economic recovery during a downturn through currency devaluation if necessary. In addition, the huge divergence between member states' economic models means that the euro incubates and sharpens existing trade imbalances, consolidating the power of the German-centred 'core' and systematically eroding the social fabric of 'peripheral' member states in the process. Saving the EU, on this understanding, means acknowledging the iniquitous structure of EMU and returning monetary sovereignty to member states.

Those arguing for the abolition of the euro often point to the political constraints which render federalist solutions unrealistic. However, the feasibility of abandoning the euro is itself highly questionable. A recent Eurobarometer survey shows that–despite its many dysfunctions–support for the single currency remains high. In March 2017, 70 per cent of respondents cited their support for the single currency (Romei 2017). Even in Greece, where euro membership has delivered over half a decade of punishing conditionality and devastatingly high levels of unemployment, large swathes of the Greek population favour continued euro membership over a return to the Drachma. Devaluation would greatly increase import prices, constraining the spending power of Greek consumers and liquidating huge sums of savings denominated in euros. As such, the 'nuclear' option of euro abolition itself looks unlikely to command popular support.

The result is that the eurozone is caught in a seemingly intractable political and economic trap. Supranational federalist solutions aimed at large scale redistribution through the EU institutions reveal themselves to be unfeasible once one moves from the sphere of ideal political theory to the terrain of domestic politics and political economy. The abolition of the euro may appeal to some sections of the nationalist right as well as to critics of the single currency amongst the left intelligentsia, but this fix is highly unlikely to command widespread support amongst European electorates, at least insofar as they are currently constituted. Instead, management of the single currency has fallen back on a conventional ordo-liberal logic, whereby fiscal discipline is entrenched through quasi-constitutional mechanisms whilst control over key economic policy levers is increasingly insulated from the capricious whims of domestic electorates. But it is unlikely, too, that this 'solution' will hold. Enduring imbalances in the eurozone–unsustainable debt burdens, an imperilled banking sector, huge levels of youth unemployment–suggest that contemporary 'fixes' are not sustainable either economically or politically in the medium term.

The notion that the eurozone is snared in an intractable dilemma has become increasingly prominent. Claus Offe has argued that Europe is 'entrapped' in the dysfunctional architecture of the single currency, whilst Magnus Ryner has argued that the eurozone embodies an 'ordo-liberal iron cage' (Offe 2015; Ryner 2015). As Offe remarks, the core of the euro's crisis is *agential*. Technical solutions may exist to the eurozone debacle, but no social or political force currently has the strength to articulate an alternative to the status quo. In the absence of such an agency emerging, the deep imbalances afflicting the eurozone will most likely persist and even intensify into the future.

In a context of slowing Chinese growth, protracted difficulties in the emerging economies and highly unpredictable developments within Anglo-America, the possibility of a second great economic downturn is by no means unthinkable. Indeed, it is looking increasingly likely. If such an event were to unfold, the imbalanced eurozone is likely to be front and centre of the global maelstrom. Monnet's dictum–that Europe will be deepened rather than overwhelmed by its responses to dysfunction and disequilibrium–could face its greatest test yet if the coming crisis does eventually materialise.

References

Becker, J., & Jäger, J. (2011). European integration in crisis: the centre-periphery divide, *Euromemo working paper.* Retrieved from http://www2.euromemorandum.eu/uploads/ws5_becker_jaeger_european_integration_in_crisis._the_centre_periphery_divide.pdf.

Draghi, M. (2014). Statement to the press conference. *European Central Bank*. Retrieved March 5, 2017, from https://www.ecb.europa.eu/press/pressconf/2014/html/is140807.en.html.

Elliot, L., & Atkinson, D. (2016). *Europe isn't working*. Yale: Yale University Press.

EuroMemo. (2015). *What future for the European Union—Stagnation and polarisation or new foundations?* Retrieved from http://www.euromemo.eu/euromemorandum/euromemorandum_2015/.

Flassbeck, H., & Lapavitsas, C. (2016). *Against the troika : Crisis and austerity in the Eurozone*. London: Verso.

Gechert, S., et al. (2015). *Fiscal multipliers in downturns and the effects of Eurozone consolidation*. Retrieved from http://www.cepr.org/sites/default/files/policy_insights/PolicyInsight79.pdf.

Habermas, J. (2014). *The Crisis of the European Union: A Response*. London: Polity.

Heyes, J., & Hastings, T. (2015). Where now for flexicurity? Comparing post-crisis labour market policy changes in the European Union, *SPERI Global Political Economy Brief No. 3*. Retrieved from http://speri.dept.shef.ac.uk/wp-content/uploads/2016/03/Global-Brief-3-Where-now-for-flexicurity.pdf.

Khan, M. (2015, August 6). Indebted Portugal is still the problem child of the eurozone. *The Telegraph*. Retrieved from http://www.telegraph.co.uk/finance/economics/11786694/Indebted-Portugal-is-still-the-problemchild-of-the-eurozone.html.

Oberndorfer, L. (2015). Asymmetric crisis in Europe and possible futures. In J. Jäger & E. Springler (Eds.). *Asymmetric governance and possible futures: Critical Political Economy and post-Keynesian Perspectives* (pp. 186–207). Abington: Routledge.

OECD. (2014). Government at a glance 2014: Greece. *OECD Country Fact Sheet*. Retrieved March 5, 2017, from https://www.oecd.org/gov/Greece.pdf.

Offe, C. (2015). *Europe entrapped*. Cambridge: Polity.

Romei, V. (2017, March 2). Support for the euro back at record high. *The Financial Times*. Retrieved from https://www.ft.com/content/020a215e-ff28-11e6-8d8e-a5e3738f9ae4.

Ryner, M. (2015). Europe's ordoliberal iron cage: Critical political economy, the euro area crisis and its management. *Journal of European Public Policy, 22*(2), 275–294.

Scharpf, F. W. (2015). After the crash: A perspective on multilevel European democracy. *European Law Journal, 21*(3), 384–405.

Stockhammer, E. (2016). Neoliberal growth models, monetary union and the Euro crisis: A Post-Keynesian perspective. *New Political Economy, 21*(4), 365–379.

Streeck, W. (2014). *Buying time: The delayed crisis of democratic capitalism*. London: Verso.
Streeck, W. (2015). Why the euro divides Europe. *New Left Review, 95*, 5–26.
Verdun, A. (2015). A historical institutionalist explanation of the EU's responses to the euro area financial crisis. *Journal of European Public Policy, 22*(2), 219–237.

Systemic Stabilisation and a New Social Contract

Andrew Baker and Richard Murphy

Abstract We live in an era characterised by a complex and dynamic relationship between financial innovation, the state and patterns of investment. At its core is the little understood issue of shadow money—a 'promise to pay' backed by high-grade collateral, usually government bonds, which means that government debt now plays a key role in the stabilisation of the financial system. Central banks' growing appreciation of how shadow money can generate destabilising dynamics has necessitated them to take preventative actions, and new forms of central bank-led systemic stabilisation have materialised. This new and complex dynamic requires a new social contract, the investment state—a compact between the state and financial markets and between the stabilisation and investments arms of the state.

Keywords Systemic stabilisation · Social contract · Shadow money
Investment state · Central banks

A. Baker (✉)
SPERI, University of Sheffield, Sheffield, UK

R. Murphy
City University, London, UK

C. Hay and T. Hunt (eds.), *The Coming Crisis*, Building a Sustainable Political Economy: SPERI Research & Policy,
DOI 10.1007/978-3-319-63814-0_11

One of the most penetrating analyses of the global financial crash of 2008 notes that it emerged from shadow banks and the practices of shadow money creation in the United States (Pozsar et al. 2010). That the biggest crisis to rock the global economic system for over seventy years emerged from these markets, means that meaningful commentary on any coming crisis need to assess current trends in these markets, asking whether they have become more stable since 2008, or whether they continue to generate instability and systemic risks. One crucial thing has clearly changed since the crash. Central banks have a growing appreciation of how shadow money and shadow banks in 'repo markets' can generate destabilising dynamics, necessitating preventative action from them. Aside from assessing the extent to which the shadow financial world has become more stable, in this contribution we also ask the all important question of what the political implications of increased central bank activism in this domain are, and could be, in light of a potential coming crisis.

The present era is characterised by a complex and dynamic relationship between the state, financial innovation and patterns of investment that is little understood. At the centre of these relationships are the practices of shadow money which consists of 'repos'. Repos are repurchase agreements (repos), an effective 'promise to pay' between financial institutions backed by high-grade collateral, usually government bonds. The issuing of shadow money takes place in repo markets, which involve interactions between banks and cash rich institutions such as pension funds and insurance companies. The aim is to allow institutions access to funding through the issuing of repo contracts. The run on US repo markets following the collapse of Lehman Brothers has received some notable attention (Gorton and Metrick 2012) and the US Federal Reserve, has put repo transactions at \$2.2 trillion in 2015, and reverse repo transactions at \$1.8 trillion (Baklanova et al. 2015). What is often overlooked is that repo transactions have also grown in Europe and played a central role in the eurozone crisis, which became a repo market crisis, rather than just a standard sovereign debt crisis (Gabor and Ban 2016). The European Central Bank (ECB) has suggested that in cumulative (flow) terms repo transactions had reached €25 trillion annually by 2008 (ECB 2015). In 2011, European banks were still funding 66 per cent of their assets in these wholesale funding markets. Most crucially of all, around 75 per cent of repo transactions use government bonds as collateral. Repo markets are central to the funding models of many financial institutions; they connect credit and financial markets and arguably the entire financial system, to the market for government debt.

For shadow banking guru, Zoltan Pozsar (formerly of the New York Federal Reserve but now with Credit Suisse AG), a run on the repo market, equivalent to the one in 2008 is unlikely today in the USA. However, this is largely because of the actions and positions the Federal Reserve has taken. In this reading, growth in the Fed's balance sheet and less liquid capital markets has made the global financial system inherently safer. Pozsar has argued that the Fed appears to be revamping itself to become a new global funding force (Pozsar 2016). In this scenario, the Fed's extended balance sheet may prove to be a permanent state rather than a temporary phenomenon in the aftermath of years of stimulative monetary policy. Increased reserve requirements for banks are driving increased demand for the Fed's balance sheet. In the period prior to the crash of 2008, banks were allowed to increase their exposure to wholesale funding without a coinciding increase in reserves to help insulate them for the vagaries of short-term financing. Now, however, banks are increasingly holding their reserves at the Fed to meet the new liquidity coverage ratio (Pozsar 2016). One way of reading this is that larger more sophisticated financial markets, also require larger more financially active central banks.

But how did we get to where we are and what are the implications of new central bank functions? First, any appreciation of a 'coming crisis,' must begin with an acknowledgment that we have never really left the previous one behind. The collapse in asset values and dramatic evaporation of wealth of 2008 resulted in what the Bank for International Settlements (BIS) call a balance sheet recession. These are notoriously protracted affairs. Evidence of this can be found in the publicly-expressed worries of a diverse range of economists. BIS staff worry about the paralysis of monetary policy caused by debt overhang and over indebted agents (Borio 2012). Adair Turner, the former chair of the UK's Financial Services Authority, has warned of the continuing problems of global deflation and disinflation, calling on authorities to start exploring overt monetary financing (Turner 2015). Andy Haldane, the Bank of England's Chief Economist has alluded to the potential options of negative rates and the effective abolition of cash payments (Haldane 2015). Simon Wren-Lewis has made the case for a form of helicopter money (Wren-Lewis 2016). Larry Summers references secular stagnation and the need for greater combined monetary and fiscal stimulus (Summers 2016). All are extraordinary suggestions for extraordinary times. Economic, policy and intellectual elites of a variety of stripes are deeply concerned and troubled.

The common thread behind most of those analyses is demand deficiency. As Yanis Varoufakis has noted there is a shortfall in investment, particularly, in the things we need most. These include environmentally friendly sustainable new technologies, infrastructure projects and research and development, all of which suffer from progressively shortening financial market time horizons (Varoufakis 2016). The consequence is that investments in the very things that do most to generate productivity, growth and meaningful long-term work, are at risk in an age of asset management which is dominated by short-term logics (Haldane 2014).

As recent work by Daniela Gabor and Jakob Vestergaard (2016) has shown us, the operation of repos markets come with a conspicuous downside. The shadow money system is complex, but its primary relevance for the theme of a coming crisis, is its pronounced 'procyclicality' and inherent fragility. The liquidity of financial institutions' with repo funding depends on their capacity to settle obligations with immediacy—an ability to be able to convert promises into sovereign state money on demand. If financial assets in an institution's portfolio start to fall, institutions with repo liabilities need to sell assets to raise cash to meet their obligations. This can lead to fire sales of assets, but also downward liquidity spirals. The conversion of repo liabilities by their holders in a dash for cash, inadvertently exert a downward pressure on their collateral valuation (government bonds). The conversion of repo claims into sovereign state money can consequently be compared to climbing a ladder that is sinking—the faster you climb, the faster it sinks (Gabor and Vestergaard 2016, p.25).

Consequently, market liquidity has become a pivotal social institution in market-based finance, but as Keynes noted the 'illusion of liquidity' means it is notoriously fickle and prone to sudden evaporations (Keynes 1930). Lehman Brothers and Bear Stearns experienced these processes when they lost access to repo funding in 2008, making it impossible to meet repayment demands on funding agreements.

Government bonds are the most favoured form of shadow money collateral because they are on the whole liquid and low risk. Government debt accordingly plays a crucial role in the balance sheet expansion of financial institutions. Consequently, the expansion of market-based financial systems actually places new demands on the state to issue debt, because financial institutions need that base asset to support credit expansion. However, in Europe, shadow money vulnerabilities and liquidity problems have in the recent past spread to government bond markets.

In 2012, the ECB engaged in a programme of Outright Monetary Transactions (OMT) purchasing eurozone member bonds. This coincided with concerns in repo markets about the collateral grade status of Italian bonds (Gabor and Ban 2016). Plummeting Italian bond prices caused by the kind of downward liquidity spiral described above and fire sales would have obliterated collateral valuations in repo markets, taking Europe's financial system to the brink. Through OMT, the ECB essentially committed to 'whatever it takes' and in the process saved the European repo market by essentially, backstopping core shadow money markets, with the central bank placing a floor under the base asset market—government bonds. A new form of central bank-led systemic stabilisation had materialised. In the process, a monetary policy operation was implemented that had both fiscal and financial stability implications. An era of big complex money markets is therefore arguably eroding the neat prior segmentation between monetary, fiscal and financial stability. Central banks now face a tricky balancing act between stabilising the system of shadow money and maintaining base asset scarcity, while avoiding excessive government bond purchases.

If we want to look for evidence that this may not be an isolated incident but part of a broader central bank recognition of a new necessary systemic stabilisation function, we need to look no further than the Bank of England's Red Book. The relevant passage reads 'in exceptional circumstances, the Bank stands ready to act as a market maker of last resort. Any such intervention would aim to improve the liquidity of one or more markets whose illiquidity posed a threat to financial stability, or was judged to be important to the transmission mechanism of monetary policy' (Bank of England 2015, p.6). The ECB made public similar rationales when it announced OMT. Central banks recognise the instability of contemporary money markets by offering an implicit backstopping guarantee.

The recognition by central banks of their systemic stabilisation function is an indicator of how our post-crash political economy is characterised by a new age of uncertainty and instability. The shadow money system is at the core of this instability and uncertainty and the state's role in this system is already being forced to evolve by market dynamics. Economics is beginning to catch up. Party politics lags further behind. Systemic stabilisation places several important political issues on the table that need attention. The backstopping of government debt in the name of repo market stabilisation that we have seen in the UK and Europe is arguably a collective welfare enhancing function, that keeps the banking and credit creation system on the road. But this welfare enhancement and its benefits are distributed asymmetrically towards those institutions directly operating in repo markets, who essentially have their base asset protected and safeguarded at no cost to themselves.

Signalling a market backstopping function is good for market confidence, but it emphasises the need to reconsider the accountability relationship and delegation contract that central banks operate under. In particular, it is suggestive that a new emerging systemic stabilisation function for central banks ought to be accompanied by a new social contract (Gabor and Vestergaard 2016, pp.30–31). Clarifying systemic stabilisation mandates is one element. Another relates to coordination of debt issuance and management, including the appropriate relationship and channels of communication between Treasuries and central banks, as the boundaries between fiscal, monetary and financial stability policy blur. In the context of this brave new financial world, a review of the institutional design of major central banks, their functions and mandates is most certainly warranted.

Most crucial of all is the question of what those private financial institutions benefiting most directly from the stabilisation of repo collaterals should contribute in return for the public backstopping of these markets. Central banks were not set up to disproportionately aid and abet big finance. For the sake of legitimacy, there is a need to ask for something back from financial institutions in return as part of a new social contract, whether through specific repo taxes, or through an obligation placed on them that they must buy the bonds of some sort of new public investment funds. Already, we are seeing the political fallout from a perceived privileging of finance, when the population at large suffers due to stagnation and fiscal austerity, through the rise of populism with a nationalistic tinge throughout Europe and North America. Both Brexit and the election of Donald Trump are a phenomenon that pose threats to financial stability, because within the politics of both there is an impetus for a further round of financial deregulation. Central banks are fearful that financial systems are not sufficiently robust and resilient, following the last crash, to withstand the risk taking this would bring.

A new social contract tying financial markets and their actors into some sense of collective social obligation could not only restore much-needed legitimacy but create a shared pool of capital, which could be the basis for the operation of new state investment arms to accompany their stabilisation arms. Such an investment bank with a broad capital base, including private stakeholders, could invest in infrastructure projects that could stoke demand in a countercyclical fashion while creating private financial incentives and interest in the success of such projects. More important would be the political symbolism associated with such moves. This would signal a social contract suggestive of a greater degree of social balance, fairness and compromise, rather than one in which big corporate actors get most of what they want, most of the time, but offer little in return.

Irrespective of precise institutional details, the question of what a new social contract for central banks' new systemic stabilisation function should look like needs to be asked. It is, of course, a bridge too far to ask central banks to start talking about, less still designing, new social contracts. But legislatures, political parties and non-governmental organisations need to step up to the mark in making the case. Governments could usefully convene expert panels to ask these questions and to report, with concerns of legitimacy, fairness and widespread public consultation, paramount.

The precise nature and terms of this new social contract, and how it should inform institutional design is central to the political economy of the new era of great uncertainty. Any coming crisis would throw a spotlight once again on the shadow money system—and more generally on the practices of the financial sector and their social and economic consequences. We need to build a workable politics for the investment state and the new social bargains that will be needed to tie a variety of stake holders into its operation, and we need to do so now rather than waiting for the next crisis to come.

References

Baklanova, V., Copeland, A., & McCaughrin, R. (2015). *Reference guide to US repo and securities lending markets* (Federal Reserve Bank of New York Staff Report No. 740). https://www.newyorkfed.org/medialibrary/media/research/staff_reports/sr740.pdf.

Bank of England. (2015). *The Bank of England's sterling monetary framework.* Updated July 2015.

Borio, C. (2012). *The financial cycle and macroeconomics: What have we learnt?* (BIS Working Paper, no. 395).

ECB. (2015). *Financial integration in Europe, 2015.* Available at https://www.ecb.europa.eu/pub/pdf/other/financialintegrationineurope201504.en.pdf.

Gabor, D., & Ban, C. (2016). Banking on bonds: The new links between states and markets. *JCMS. Journal of Common Market Studies, 54*(3), 617–635.

Gabor, D., & Vestergaard, J. (2016). *Towards a theory of shadow money* (INET Working Paper). https://www.ineteconomics.org/uploads/papers/Towards_Theory_Shadow_Money_GV_INET.pdf.

Gorton, G., & Metrick, A. (2012). Securitized banking and the run on repo. *Journal of Financial Economics, 104*(3), 425–451.

Haldane, A. G. (2014, April 4). *The age of asset management?* Speech at the London Business School. http://www.bankofengland.co.uk/publications/Documents/speeches/2014/speech723.pdf.

Haldane, A. (2015). *How low can you go.* Speech given by chief economist of the Bank of England at the Portadown Chamber of Commerce.

Keynes, J. M. (1930). *A treatise on money.* London: MacMillan.

Pozsar, Z., Adrian, T., Ashcraft, A., & Boesky, H. (2010). *Shadow banking* (Federal Reserve Bank of New York Staff Reports No. 458).

Pozsar, Z. (2016, April 13). *What excess reserves? Global Money Notes #5.* Zürich: Credit Suisse. https://research-doc.credit-suisse.com/docView?sourceid=em&document_id=x682509&serialid=XzV6W6kii66%2B6SsdSZ0rpA2M1jUfVJkwoq9GntIjPBc%3D.

Turner, A. (2015). The case for monetary finance—An essentially political issue. In *16th Jacques Polak IMF, Annual Research Conference.*

Summers, L. H. (2016). The age of secular stagnation: What it is and what to do about it. *Foreign Affairs, 95,* 2.

Varoufakis, Y. (2016, March 2). My advice to UK's Jeremy Corbyn and George Osbourne. *Newsweek.* http://www.newsweek.com/brexit-george-osborne-jeremy-corbyn-yanis-varoufakis-432637.

Wren-Lewis, S. (2016). *Helicopter money and fiscal policy.* https://mainlymacro.blogspot.co.uk/2016/05/helicopter-money-and-fiscal-policy.html.

Secular Stagnation: The New Normal for the UK?

Jonathan Perraton

Abstract A range of authorities has argued recently that rather than the 2007–08 global financial crisis representing a major, but one-off, downturn, mature economies have now entered a period of 'secular stagnation' where underlying productivity growth rates have fallen to very low rates. Growth has only been sustained by various bubbles and is now likely to slump to low rates for the foreseeable future. Debates have largely focused on the US but the concept of secular stagnation in the context of UK requires examination. Before the crisis the UK economy saw particularly strong property price booms and credit expansion, with demand sustained by what has been termed 'privatized Keynesianism'; the post-crisis UK economy is now experiencing prolonged stagnation in productivity and growth again appears to be driven by debt-financed private consumption.

Keywords Secular stagnation · Productivity · Growth · Debt · UK

J. Perraton (✉)
University of Sheffield, Sheffield, UK

C. Hay and T. Hunt (eds.), *The Coming Crisis*, Building a Sustainable Political Economy: SPERI Research & Policy,
DOI 10.1007/978-3-319-63814-0_12

It is now more than eight years since the collapse of Lehman Brothers heralded the Global Financial Crisis (GFC), roughly the length of a normal business cycle. Over that time interest rates have been reduced to record low levels and the Bank of England has augmented this with a major programme of quantitative easing. Yet economic recovery in the UK has been slow and remains fragile and anaemic, whilst productivity has stagnated. Real wages face a decade of stagnation; as Bank of England Governor Mark Carney noted such a prolonged period of wage stagnation has not been experienced in the UK since the 1860s (Carney 2016). These trends were evident even before the Brexit vote; since then growth forecasts have been revised downwards despite interest rates being lowered further still.

Recently, several leading economists have revived older notions of secular stagnation, the inability of developed countries to resume past growth rates even with ultra-loose monetary policies. Recovery from the GFC has not simply been a case of fixing the financial system; instead, there is the prospect of slow growth for the foreseeable future. There are variants of this argument and use of the term varies between authors. Former US Treasury Secretary Larry Summers has argued that developed economies have found it increasingly difficult to achieve adequate demand growth and financial stability. Periods of expansion, in his argument, have been limited and based on the bubbles in housing and other asset markets and associated with unsustainable growth in debt. Meanwhile, private investment since the crash has remained low and companies have accumulated cash surpluses rather than borrowing to expand. Even during the 'Great Moderation' period before the GFC, investment activity was not especially high; the falling real interest rates during the 2000s failed to stimulate private investment.

Long run estimates of the interest rate consistent with normal levels of output and employment have trended downwards for the US, UK and other major economies; although there are major conceptual and empirical question marks over these estimates, they do point to the limits of relying on monetary policy to maintain demand in the economy. Since the GFC the zero lower bound has limited central banks' ability to boost demand and, in any case, the resulting low interest rates may undermine financial stability again in the medium term. As developed economies have struggled to generate sufficient levels of demand, Summers argues that policies brought in as emergency responses to the GFC are likely to persist, but without supporting fiscal policy they will be insufficient to generate pre-crisis growth rates (Summers 2014). A prolonged shortfall in demand would damage future growth prospects by reducing incentives for investment. Some US commentators have further located these trends in terms of rising inequality–as wages stagnated, households have attempted to maintain their consumption levels through lower saving and higher borrowing; stagnation in demand was only held off by household borrowing that ultimately proved to be unsustainable (Palley 2012).

Robert Gordon—a long-standing sceptic of the economic growth potential of new information and communication technologies (ICTs)—takes an even longer term view. In *The Rise and Fall of American Growth*, Gordon argues that there has been a slowdown in long-run growth rates so that the improvements in living standards the US enjoyed in the twentieth century will not be sustained in the twenty-first (Gordon 2016). The key innovations that sustained rising living standards have petered out and a series of headwinds are likely to hold back growth in coming decades. This is largely due to a slowdown in the underlying rate of technical progress, but also reflects the impact of globalisation on wages for much of the workforce.

Much of the recent discussion of secular stagnation has focused on the US experience, but there are a number of key similarities with the UK case as well as some key differences. As Jeremy Green notes in his chapter, contemporary discussions of secular stagnation should be seen in the context of longer term structural weaknesses of the UK and developed economies since the end of the post-war Golden Age. Whilst the UK's experience of sluggish recovery from the GFC is hardly unique, key features of the UK growth model raise the possibility that it now faces secular stagnation too. The UK model has become centred around the growth of household consumption spending. The Great Moderation saw growth sustained by private consumption through a housing boom and associated fall in household savings. As a result, outstanding household debt relative to income soared in the 2000s to levels that were high both by historic standards and relative to other countries. Although households consolidated in the initial aftermath of the GFC and built up their savings, in the current decade with the resumption of house price rises, savings have fallen to historic lows relative to household income and private household credit has grown rapidly once again.

Overall, the household debt-to-income ratio has risen since 2012 to levels approaching their pre-GFC peak. The trends in the 2010s have meant that whereas this ratio fell initially during the GFC there has been minimal consolidation since and household debt remains at higher levels than in many other developed economies. The era of very low interest rates since the start of the GFC has kept borrowing costs low, but households remain vulnerable to even a modest rise in interest rates. Nevertheless, the Office for Budget Responsibility forecasts that household debt will continue to rise over this decade in the face of a squeeze on living standards (OBR 2016). The UK has experienced its slowest recovery from a recession for over a century and again the modest growth we have seen has been largely driven by private consumption supported by falling savings and higher borrowing. By contrast, private investment has remained low, whilst the corporate sector has been in surplus for long periods since 2002; non-financial companies are saving, rather than borrowing to expand. Exact historical comparisons are problematic for data reasons, but the UK's reliance on private consumption growth appears to be unprecedented. In the post-war period, the UK did often have relatively low private investment rates and periodic expansions based on rising consumer demand, but the degree to which demand has become based on rising household expenditure and the impact of this on falling household savings and rising debt, appears to be unprecedented.

These trends also have their impact on the 'other deficit'—the UK's external current account deficit. Again the post-war period did see regular episodes where a consumer boom in the UK led to a worsening balance of payments; however, since the GFC the current account deficit has hit a record peacetime deficit and whilst the causes of this are complex and go beyond just rising private consumption, this too raises questions over the sustainability of current economic trends. Even before Brexit the Office for Budget Responsibility was predicting that the UK's share of world exports would continue to fall; higher barriers to trade with its largest and closest market are only likely to worsen this. Sterling has fallen since the Brexit vote, and the resulting higher import prices are projected to squeeze living standards, but this is unlikely to be sufficient to restore current account balance.

Further, the longer term picture points to weak or stagnant UK productivity growth, which has been particularly weak since the GFC, with only Japan amongst the G7 countries currently displaying lower output per hour worked. This prolonged stagnation of productivity is unprecedented in the post-war period. The UK 'productivity puzzle' is compounded here, since the pre-crisis productivity growth appeared to be relatively strong, based on the application of ICTs and improvements in human capital. The underlying regulatory framework and provision of

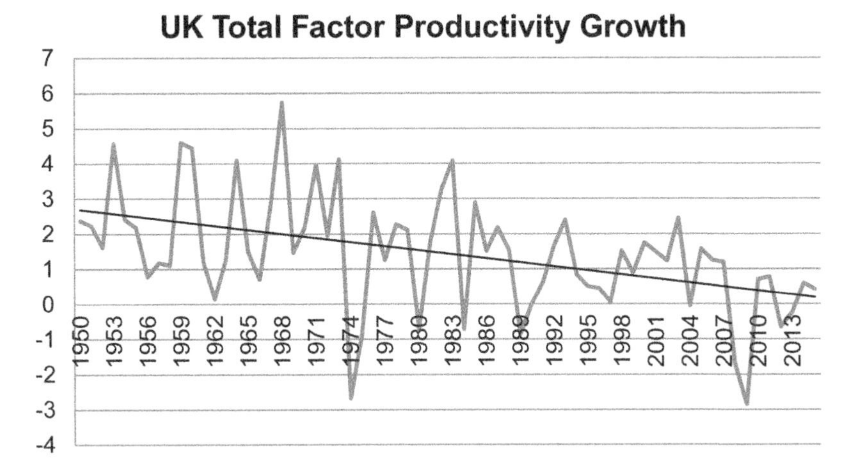

Fig. 1 UK Total Factor Productivity Growth from 1950 (%) (author's own calculation from the Bank of England's 'three centuries of macroeconomic data' dataset)

human capital have not changed substantially but whereas the UK partly closed its productivity gap during the Great Moderation, since then it has slipped back to the extent it is now around 18 percentage points below the G7 average, a position not seen since 1991. Taking a longer term view, UK productivity growth has declined over the post-war period, with productivity performance particularly weak since the GFC (Fig. 1).

Central to this is low private investment. Sustained expansion needs to be based on investment; productivity growth typically requires investment in new capacity. Sluggish investment since the onset of the GFC is a general phenomenon, consistent with explanations in terms of aggregate demand deficiency, but UK investment levels are particularly low. This can, though, be seen partly in terms of austerity and aggregate demand deficiency. In particular, Martin and Rowthorn (2012) have shown in detail that the UK economy has suffered from aggregate demand deficiency since the onset of the GFC and propose that it is this, rather than widely proposed supply-side explanations, which accounts for much of the productivity slump.

There are other factors at work here too. Although the UK economy has clearly experienced austerity since the GFC, so too have other industrialised countries (particularly, in the eurozone). Between 2008

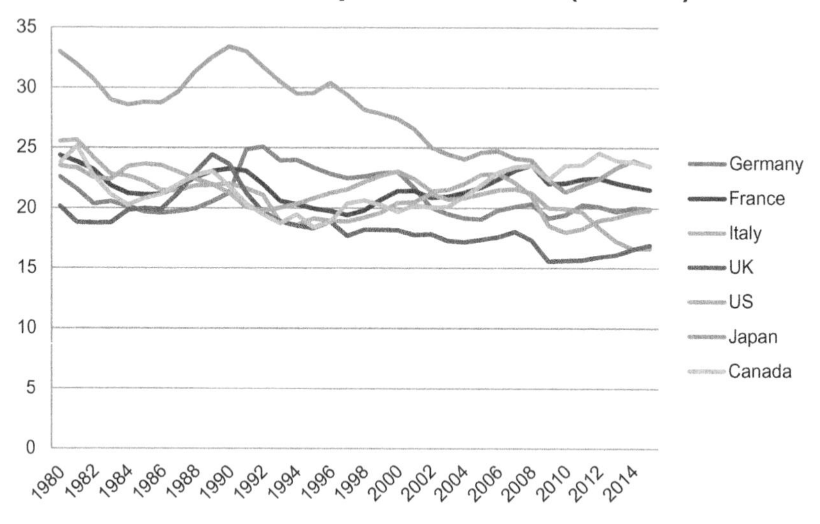

Fig. 2 Total investment, G7 economies (author's calculations from AMECO database)

and 2015 the UK saw the largest fall in real wages of any industrialised country besides Greece (ILO 2016). In large measure, this stagnation reflects the productivity slump, but it has also created a situation where the combination of weak demand and low wages provides little incentive for capital investment. Weak investment, though, undermines the basis for future growth. Over the longer term, UK investment has been particularly weak, typically having the lowest rate amongst major industrialised economies (Fig. 2).

The UK has seen a secular decline in the propensity for firms to invest out of their retained profits. There are a variety of causes of this, particularly related to the financialisation of the UK economy. Technology optimists prophesise that a range of new innovations will lead to faster future growth and claim that the impact of ICT investments on productivity has yet to be fully realised. Gordon (2016) provides detailed grounds for scepticism here. The economic potential of many of these technologies is unclear, as is the ability of firms to realise returns on investment in them. The UK has already seen substantial ICT investment and evidence from the labour market points to its impact having already been realised from the 1990s.

There are few tangible signs that a wave of innovations is likely to stimulate a sustained rise in investment. A further headwind facing the UK economy is the rising cost of health care for an ageing population. The relationship between demographic trends and economic growth is complex here; latterly Britain has been unusual amongst developed economies in experiencing sustained population growth, including growth of the labour force. Much of this, though, has been driven by immigration and following Brexit, this may not continue.

Overall, developments in the UK mirror much of the secular stagnation story for the US economy over the longer term. The period of sluggish recovery can be seen as reverting to the prior UK growth model of house price bubbles and an associated 'privatised Keynesianism' credit boom (Crouch 2009; Hay 2011). Such booms are typically associated with lower subsequent growth and worsening current account positions. Productivity growth has stagnated. There is little sign of a sustained recovery in private investment, let alone in net exports, that could reverse this. The UK has a long-standing tendency for relatively low investment, with corporate cash surpluses and a secular decline in the propensity for firms to invest out of their retained profits. This is not a sustainable basis for long-term growth. Macroeconomic policy has effectively been reduced to monetary policy, but even maintaining interest rates at historic low levels has been insufficient to stimulate investment and growth. As Jeremy Green notes in his chapter, the constraints on economic policy options are as much political as economic notwithstanding Summers and others advocating a major expansion of public investment as a response to stagnation. Brexit is only likely to aggravate the trends noted here by squeezing living standards, worsening the investment climate and further weakening the UK's trade position. The attractiveness of the UK for inward investment will be substantially weakened. Consensus forecasts point to clear losses in income; trade and inward investment tend to raise productivity through a variety of channels and Brexit is therefore also likely to further weaken underlying productivity growth. Without a reorientation of the economic policy framework and the underlying growth model, the UK may be facing a 'new normal' of sluggish growth, with periodic short-lived consumption booms, for the foreseeable future.

References

Carney, M. (2016). The spectre of monetarism. *Roscoe Lecture*. Available at http://www.bankofengland.co.uk/publications/Documents/speeches/2016/speech946.pdf.

Crouch, C. (2009). Privatised keynesianism: An unacknowledged policy regime. *British Journal of Politics and International Relations, 11*(3), 382–399.

Gordon, R. J. (2016). *The rise and fall of American growth: The U.S. standard of living since the civil war*. Princeton, NJ: Princeton University Press.

Hay, C. (2011). Pathology without crisis? The strange demise of the Anglo-liberal growth model. *Government and Opposition, 46*, 1–31.

ILO. (2016). *Global wage report 2016/17: Wage inequality in the workplace*. Geneva: International Labour Organization.

Martin, B., & Rowthorn, R. (2012). *Is the British economy supply constrained II? A renewed critique of productivity pessimism*. Cambridge: University of Cambridge.

OBR. (2016, November). *Economic and fiscal outlook*. Cm 9346.

Palley, T. (2012). *From financial crisis to stagnation: The destruction of shared prosperity and the role of economics*. Cambridge: Cambridge University Press.

Summers, L. (2014). U.S. economic prospects: Secular stagnation, hysteresis, and the zero lower bound. *Business Economics, 49*, 65–73.

China Crisis?

Matthew Bishop

Abstract Since the onset of China's massive industrial expansion in the 1980s, predictions of a coming crisis have consistently been proven wrong as its growth trajectory has continued, decade after decade, seemingly unabated. But are things different this time? To answer this question, the contemporary moment must be set within the wider context of China's substantial developmental transformation. A range of challenges lend weight to the notion of an impending crisis, but there is significant uncertainty about the likely extent of adjustment that will be required. How Chinese policymakers grapple with these imperatives, and the way that their costs and consequences are distributed, will ultimately shape the next stage of China's spectacular development process.

Keywords China · Growth · Development · Globalisation
Transformation

China has apparently been on the cusp of a crisis for decades. In the 1990s, as its nascent industrialisation gathered momentum, neoliberal economists regularly advocated 'shock therapy' of the kind undertaken

M. Bishop (✉)
University of Sheffield, Sheffield, UK

C. Hay and T. Hunt (eds.), *The Coming Crisis*, Building a Sustainable Political Economy: SPERI Research & Policy,
DOI 10.1007/978-3-319-63814-0_13

disastrously by Russia as a way of staving it off (Arrighi 2007, p.15). In the 2000s, Western observers rehearsed familiar arguments about the assumed incompatibility between authoritarian government and market reform. That the crisis never came should not surprise us: the overwhelmingly ideological criticisms levelled at the Chinese model of political economy over the years have been predicated on stylised assumptions about what constitutes a 'market' or a 'democracy' distilled from often-misguided, teleological interpretations of the supposedly superior Western experience (Bishop 2016a). As we reach the late 2010s, though, after four decades of astonishing economic growth, the question of a coming crisis has reared its head once more. Yet the past is not always a good guide to the future: just because the critics have been repeatedly wrong, it does not follow that they are this time. Growth in China is slowing, its burgeoning imbalances are increasingly evident, and question marks hang over the future.

The Chinese Transformation

In 1990, China was a low-income country with GDP per capita of around US $300, on a par with the poorest in sub-Saharan Africa. Today, it is rapidly approaching US $8000 and is projected to reach US $11,500 by 2020, higher than all of the other so-called 'BRICS' countries (Brazil, Russia, India and South Africa). These are staggering figures, particularly so in a country with a population of 1.3 billion, where tens (even hundreds) of millions still eke out a subsistence living in the countryside. In the major metropolitan regions, average incomes are now hovering around US $20,000, and given that the Renminbi has long been undervalued, in purchasing power parity (PPP) terms the Chinese are actually considerably wealthier than we may often realise. Indeed, the economy may even be larger than the US equivalent on a PPP basis.

To get a sense of the scale of change, it is worth noting that China's growth rate has not only been positive for every year in the past forty or more, but has generally hovered between 7 and 15 per cent since the 1990s. The economy has therefore grown exponentially and consistently, and has more than *doubled* in size over the past decade alone. Other facts emphasise the point: despite often being considered together in analyses as the key emerging countries, China's economy is *five times* the size of its Indian equivalent, and the latter's population is only slightly smaller.

Yet, growth is not only important in and of itself. Plenty of countries grow, often quite rapidly, but their economic panorama does not substantially change. The rapid expansion of primary and secondary sectors of the economy can actually inhibit meaningful industrialisation and diversification. The end of the commodity boom is making this abundantly clear in many Latin American countries that remain dependent on a relatively narrow range of exports. This is something that may ironically have been encouraged by China's thirst for resources since the turn of the millennium, which has both helped to lock others into commodity production while also assisting China in out-competing them in global export markets for manufactures (Gallagher and Porzecanski 2010).

By contrast, enduring growth in China has been accompanied by a fundamental transformation of the scale and scope of its productive capacity, such that the country has effectively reached an altogether different plane of developmental possibility (Bishop 2016b). Unlike other high-growth 'emerging' economies, China has not simply seen a boom in a narrow range of commodity sectors. It has rather engaged in successive processes of industrial upgrading, shifting the nature and parameters of its patterns of comparative advantage, and ultimately undergoing the kind of authentic industrial revolution that occurred in Britain in the late eighteenth century, Germany and the US in the early nineteenth, and the East Asian 'Tigers' in the 1970s (see Chang 2002).

This manifests itself in different respects. First, there is the simple visceral sense that China today *feels like* a 'developed' country in ways that it did not in the past (and others still do not). This is clearly evident in the construction of infrastructure: China has laid more high-speed rail in a decade than the whole of the rest of the world has done in half a century. Moreover, dozens of cities are simultaneously building metro lines, each a similar size to the London Underground, from scratch.

Most importantly, the Chinese economy today is not what it was. While its early expansion was built on leveraging a comparative advantage in cheap labour and textiles, clever policies to facilitate investment and ensure technology transfer—including joint public/private ventures between domestic and foreign capital—have seen the gradual, systematic upgrading of production (Lin 2014). Firms are now increasingly outsourcing low value-added activities to countries elsewhere in Asia and Africa (Kaplinsky 2013), and the major growth sectors are in industries at the technological cutting edge which have huge export potential: e.g. high-speed rail, renewable energy and other green technology (Mathews and Tan 2015), advanced petrochemicals and ICT.

Consequently, the economy is today both remarkably diversified, and many of the most influential multinationals, in a range of global sectors, are increasingly Chinese. They are frequently highly capitalised, and able to draw on huge reservoirs of public and private finance, given the complex inter-linkages between China's banks, sovereign wealth funds, and national and provincial state agencies. To get a sense of what this means for future investment and industrial development, consider intellectual property: Chinese organisations today successfully file four times as many US patents as their Indian equivalents, and more than twice as many as the other BRICS and Mexico combined (Serrano 2016).

The Crisis Drumbeat

Despite the fact that China enjoys greater freedom of manoeuvre than most countries, its development is still ultimately tied to specific patterns of interdependence with the global economy. Just as other fast-growing countries discovered in the past, it cannot continue to grow at the same blistering pace forever, and it is not immune to the consequences of domestic upheavals or wider patterns of international economic restructuring. The big worry stems from the unprecedented nature and sheer scale of the Chinese expansion: any serious crisis could cause shockwaves of genuinely historic proportions.

There is no doubt that the combination of intersecting challenges is potentially acute. First, there are evident financial jitters. Public debt remains relatively low at roughly 40 per cent of GDP, and it is buttressed by the government's huge sovereign wealth reserves, massive surpluses of foreign currency and bond holdings. Yet the total debt stock is much larger: public debt alone may actually be more than double this, once the borrowing of provincial and other local government bodies is accounted for (see Breslin 2014). Total public and private debt has soared from 148 to at least 249 per cent of GDP over the past decade: the unavoidable flipside, perhaps, of the scale of investment. Although not intrinsically problematic, in a context where growth has slowed to under 7 per cent and asset bubbles inflated by the unorthodox forms of monetary policy deployed in most of the West since 2008 appear dangerously close to bursting, China may find itself exposed financially.

Second, there is the imperative of rebalancing, to move away from investment-and export-led growth and towards greater domestic consumption. However, this also becomes more difficult in the context of an economic slowdown, where international export markets are now shrinking, commodity prices are collapsing, and, in the Trump-era US and post-Brexit EU, are potentially subject to further protectionist pressures and even outright trade wars. Some economists also believe China may be reaching Arthur Lewis's so-called 'Turning Point' in which the glut of rural labour dries up as urbanisation peaks, putting upward pressure on wages, thereby dramatically reducing export competitiveness (Das and N'Diaye 2013).

Third, this is occurring at the exact moment when China is struggling with immense industrial oversupply. This legacy of previous overinvestment—both cause and effect of the almost self-perpetuating export-led model—is most obviously reflected in the travails of the steel industry. In mid-2016, as Tata Steel was attempting to cut 750 jobs at the UK's largest plant at Port Talbot, the Wuhan Iron and Steel Company—just one of many gigantic state-owned Chinese firms—dismissed 50,000 people overnight (Shepherd and Pooler 2016).

If such patterns are repeated across the entire economy, in parts of the country a perfect storm is brewing as huge and increasingly indebted 'zombie' firms are unable to drive economic rebalancing due to a simultaneous decline in export competitiveness and domestic demand. These problems are magnified further if we consider the international pressure—especially from the US—for Renminbi appreciation, a process with which rebalancing is inextricably linked, as well as the wider environmental challenge facing Chinese policymakers, whereby rampant industrial growth has caused horrific levels of environmental degradation.

There is, in addition, a troubling set of political concerns haunting the economic landscape. President Xi's government has centralised power and tightened its grip over the state, facilitating an anti-corruption drive against elites, while also sharpening its repressive security apparatus. Consequently, the slowdown could precipitate resistance to necessary reforms and exacerbate this tendency (Naughton 2014). But things could conceivably get messier: in a country with hundreds of protests daily, economic upheaval may intensify social unrest, making it tempting to deploy these tools more widely. Some observers wonder whether these processes could even perpetuate themselves to such an extent that a full-blown crisis of the state—and the complete unravelling of China's political economy of the past half-century—may be the sole opportunity for catharsis (see Shambaugh 2016).

A Soft Landing?

Critical Western portrayals of contemporary China—as of the China of the past (see Hobson 2004)—are often laced with *Schadenfreude*. Past success offers no guarantee of future trajectory, but an alternative view might nevertheless be more sanguine. Growth is necessarily lower than previously, but still far outpacing any European country. The government holds foreign assets worth around half of GDP, and more foreign currency than the other BRICS combined. Debt remains a worry: *grands projets* such as high-speed rail continue to accumulate mountains of liabilities. However, much of the debt cascading through China's major institutions, firms and households is the result of stimulus injected since 2008, which has maintained growth and even helped keep the global economy afloat. This has inevitably precipitated real estate and asset bubbles, and it intrinsically renders deleveraging, and therefore rebalancing, more difficult.

Yet in contrast to post-crisis stimuli elsewhere, it has also propelled continued investment in real infrastructure and extensive—albeit perhaps excessive—industrial capacity, which in turn underpins future growth. What may appear today overleveraged oversupply, may not tomorrow. As Barry Naughton (2010, p.449) has put it: 'project after project that seemed at inception to be superfluous and wasteful now hums along as part of China's booming economy'. The effects, moreover, cannot just be measured in numbers. People in many provincial cities can travel to Shanghai or Beijing in 5 or 6 hours; previously it took a day. This has stimulated economic activity by bringing a large and historically fragmented country—its people, firms and production networks—more tightly together.

The overarching picture, then, is mixed. China faces challenges, but despite any immediate crisis appears relatively well placed to ride out the storm over the long-term because of economic fundamentals. Although the overcapacity problem is real, it is, for example a demonstrably different problem, with entirely different abiding consequences, to an over-dependence on primary commodities. This challenge—which is often obscured during boom times—has afflicted many other fast-growing developing countries that have not managed to translate growth into an analogous, wide-ranging transition marked by genuine industrial transformation (Bishop 2016b).

By contrast, the R&D of Chinese firms and research institutions is creating colossal industrial capacity at the innovation frontier, underpinning further growth, and shifting China's broader development patterns as it rapidly ascends global value chains. It may export less in tandem with rebalancing, but what it does export—e.g. green technology or high-speed rail systems—will be far more valuable than the textiles or manufactures it exported previously, and often paid for by others with loans from Chinese banks. Even were the growth rate to drop to 5 per cent, the economy would still double in size before 2030.

Politics in China is also frequently misunderstood in the West. There is public satisfaction about Xi's anti-corruption drive and his perceived strength when attacking vested interests. The country is far from a tottering kleptocracy. Therefore, the slowing pace of reform should be perhaps be viewed as the institutional maturation of a political and economic system that is considerably more rational, predictable and meritocratic than is recognised outside: CCP cadres endure years of ferocious competition to make it into the senior ranks of the bureaucracy, and when they arrive they tend to govern pragmatically (Naughton 2014).

Challenges and Opportunities Ahead

It is fairly clear that the Chinese economy is slowing down, there are significant dislocations already taking place as firms carrying intractable debts both lay off staff and mothball production, and shrinking export markets exacerbate these problems while making long-overdue rebalancing more difficult. Does this amount to a crisis, and does it even imply that China's political economy is about to completely unravel?

This seems unlikely: economic crises come at fairly regular intervals for all societies—even though they are rarely anticipated (Roubini and Mihm 2011)—and we would never expect that one such crisis, however protracted, could change the broad trajectory of Western progress. So, why would we expect that to be the case in China, other than on the basis of an ingrained Eurocentrism? It is far more plausible that the challenges the country faces, if they do constitute such a crisis, contain within them the seeds of their broader resolution. Because China's development over the past half-century is so substantial, it is hard to envisage a truly fundamental catastrophe. Of course, people thought the same of the Soviet Union in 1988, but if we hold aloft the possibility of political, economic and social disintegration, we should surely think the unthinkable about the US or Europe too.

China is a country *sui generis*. The way in which it does ultimately deal with the fallout from the current crisis—if that is what it is—will carry broader repercussions for three interlinked concerns. First, the question of the relevance of the 'Chinese model', and the extent to which it is both possible and desirable at last to break the stifling neoliberal consensus on market-driven forms of development policy and practice. Second, because China's domestic development has depended on extensive free trade, it is increasingly clear that it is up to Beijing—both for its own interests and those of the developing world that it purports to represent as a whole—to take a leading role in maintaining multilateral openness. Finally, Chinese elites need to square the circle and tell a compelling story about how to marry highly successful interventionism with a commitment to the globalisation of which Xi now appears the primary defender. How he treads the fine lines and strikes the delicate balances between the many contradictory pressures—e.g. free trade *vs.* protectionism, the security implications of China's continued rise and US relative decline, the appropriate mix of state and market in generating growth—in a context where the new occupant of the White House is capricious, erratic and apparently welcoming of conflict, will ultimately determine not only how China emerges from the coming crisis, but also the fundamental nature of the global post-crisis settlement.

References

Arrighi, G. (2007). *Adam Smith in Beijing: Lineages of the twenty-first century.* London: Verso.

Bishop, M. L. (2016a). Democracy and development: A relationship of harmony or tension? In J. Grugel & D. Hammett (Eds.), *The palgrave handbook of international development* (pp. 77–98). London: Palgrave Macmillan.

Bishop, M. L. (2016b). *Rethinking the political economy of development beyond "The Rise of the BRICS"* (Sheffield Political Economy Research Institute Paper No. 30). Retrieved from http://speri.dept.shef.ac.uk/wp-content/uploads/2016/07/Beyond-the-Rise-of-the-BRICS.pdf.

Breslin, S. (2014). Financial transitions in the PRC: Banking on the state? *Third World Quarterly, 35*(6), 996–1013.

Chang, H.-J. (2002). *Kicking away the ladder*. London: Anthem Press.

Das, M., & N'Diaye, P. M. (2013). *Chronicle of a decline foretold: Has China reached the Lewis turning point?* (IMF Working Paper No. 13/26). Washington, DC: International Monetary Fund. Retrieved from: https://www.imf.org/external/pubs/ft/wp/2013/wp1326.pdf.

Gallagher, K., & Porzecanski, R. (2010). *The dragon in the room: China and the future of Latin American industrialization*. Stanford, CA: Stanford University Press.

Hobson, J. M. (2004). *The Eastern origins of Western civilisation*. Cambridge: Cambridge University Press.

Kaplinsky, R. (2013). What contribution can China make to inclusive growth in Sub-Saharan Africa? *Development and Change, 44*(6), 1295–1316.

Lin, J. Y. (2014). *The quest for prosperity: How developing economies can take off*. Princeton, NJ: Princeton University Press.

Mathews, J. A., & Tan, H. (2015). *China's renewable energy revolution*. London: Palgrave Macmillan.

Naughton, B. (2010). China's distinctive system: Can it be a model for others? *Journal of Contemporary China, 19*(65), 437–460.

Naughton, B. (2014). China's economy: Complacency, crisis & the challenge of reform. *Daedalus, 143*(2), 14–25.

Roubini, N., & Mihm, S. (2011). *Crisis economics: A crash course in the future of finance*. London: Penguin.

Serrano, O. (2016). China and India's insertion in the intellectual property rights regime: Sustaining or disrupting the rules? *New Political Economy, 21*(4), 343–364.

Shambaugh, D. L. (2016). *China's future*. Cambridge: Polity.

Shepherd, C., & Pooler, M. (2016, April 1). China's steelmakers face own trouble at mill. *Financial Times*. Retrieved from https://www.ft.com/content/6c7b888a-f7e9-11e5-96db-fc683b5e52db.

Conclusion: The Crisis Gets Political

Andrew Gamble

Abstract The financial crisis of 2008 set in train a course of profound shocks that have shaped this post-crisis decade. The crisis has had three phases up to now, and three different epicentres, and is now entering into a fourth phase. In the three previous phases, the crisis was managed and contained but not resolved. In 2016, many observers and agencies warned that the crisis was far from over and that even greater shocks might lie ahead. The vote for Brexit in the UK and the election of Donald Trump in the US mark the opening of a fourth phase of the crisis, which is directly political in challenging the assumptions of neoliberalism and globalisation on which the recent governance of the international market order have been based.

Keywords Post-crisis · Neoliberalism · Globalisation · Brexit · Trump

Nine years have now passed since the collapse of Lehman Brothers on 15th September 2008. The collapse came after more than a year of growing signs of trouble in the international financial system, as the supply of credit became constricted, and a succession of financial institutions

A. Gamble (✉)
SPERI, University of Sheffield, Sheffield, UK

C. Hay and T. Hunt (eds.), *The Coming Crisis*, Building a Sustainable Political Economy: SPERI Research & Policy,
DOI 10.1007/978-3-319-63814-0_14

found themselves in difficulty. Governments dealt with each bank collapse pragmatically, seeking to minimise and contain the effects in order to keep the system afloat. But the costs of doing so were steadily mounting, and the US authorities decided to make an example of Lehman Brothers, one of the most highly leveraged and aggressive investment banks on Wall Street. They refused to bail it out. This sent a mountain of bad debts cascading through the system, potentially imperilling every other leading player, and threatening a meltdown of the entire financial system and the prospect of collateral damage spreading to encompass the whole economy. A slump as great as the one which followed the 1929 crash seemed possible.

That outcome was averted by swift and coordinated action by governments and central banks. The actions they took then and subsequently to sustain the system were strongly criticised across the political spectrum. Fiscal conservatives and fiscal socialists, as well as economic libertarians, objected to the banks being bailed out by the state and given the message that they were too big and too important to fail. But these criticisms were ignored. The pragmatists won the argument, although at a cost. Their critics pointed out that shoring up the system disguised deep structural problems which if not addressed would mean that the crisis would not be resolved but only postponed. Far better to let the crisis do its work of creative destruction, sweeping away many of the old structures and rigidities, allowing new experiments and new initiatives to emerge. The pragmatists contended that the key task was to stop the system breaking down and then to reform it gradually, removing the possibilities for a further breakdown and making possible an eventual return to a renewed era of prosperity and growth.

The question we now face nine years on is whether the optimists or the pessimists were right. Have political and economic developments since 2008 moved us closer to that new era of prosperity and growth or are the underlying structural problems about to reassert themselves once again? As Colin Hay and Tom Hunt argue there is certainly plenty to concern the pessimist. The chapters in this book offer a range of perspectives on these issues. What they all agree is that the political economy problems facing us are complex, intractable, and in many cases deepening.

One of the reasons for this is that recovery from the 2008 crash proved to be so slow and uneven. It has been quite unlike any previous recovery from a recession in the last seventy years. The failure of the economy to respond to any of the policy remedies which have been tried explains why, as Tony Payne points out, all the leading international institutions—the World Bank, the IMF, and the OECD—began issuing increasingly sombre warnings during 2016 about the risks ahead. The international outlook appears not to be lightening but darkening. Governments are no longer sure what to do, since doing the things which they are familiar with no longer seems to bring results. Since 2008 governments and central banks have used every possible instrument at their disposal, and the cupboard is now bare. Zero interest rate policy and quantitative easing, in particular, were used to inject liquidity into the system and keep asset prices up. But as Helen Thompson explains, although they were only expected to be temporary expedients to maintain financial stability, every attempt by policymakers to wean their economies off them has so far failed, because growth remains so weak and debt so high. US debt stands at $60 trillion, 27 times higher than in the 1970s. The austerity programmes which many governments have pursued have not succeeded in removing either debt or deficits in the system, because the political costs of doing so remain so high. Economic growth and international trade have also not rebounded and the costs of this are experienced most intensely not within the rich capitalist democracies but elsewhere in the international economy. Genevieve LeBaron points to rising inequality, with 200 million unemployed, and 75 per cent of the global workforce in temporary, short term, informal or unpaid work. Low pay, exploitation and wage theft are rife, and 21 million workers in global supply chains are estimated to be victims of forced labour.

In macroeconomic terms, the period since the crash has been notable for strong deflationary trends, in marked contrast to the 1970s when the problem was accelerating inflation. Now several countries, such as Japan, are desperately trying to push prices up. We have entered a new kind of stagflation, which some have called a new secular stagnation, something which economists were worried about in the 1930s and 1940s. Jonathan Perraton addresses the issue of whether this new secular stagnation is caused by supply or demand factors. Is it the result of a vanishing of investment opportunities despite the ever increasing pace of technological change, or is it the result of growing distributional inequality and the falling labour share which has depressed wages and boosted returns to capital? There has been much debate about this, and both sides have had powerful advocates. What makes it hard for governments is that there is no policy to deal with either problem which has much record of success. This explains why some governments have begun cautiously to move beyond current orthodoxies, and are tentatively intervening again in markets. Industrial strategies have become fashionable again, as mantras about markets always being the solution have lost some of their shine. The rethinking has even extended to finance. Jeremy Green and Jacqueline Best in their chapters explore some of the novel ideas that have been floated for increasing investment and productivity and for overcoming the deficiency of demand. They involve the blurring of the line between monetary and fiscal policy, using helicopter money to finance expenditure and consumption directly, increasing wages through direct state credits—an incomes policy in reverse. Andrew Baker and Richard Murphy argue that what is most needed is a political commitment to an investment state. The financial markets would have to agree to fund public investment projects in return for the public backstopping by the government which keeps them liquid and solvent. Yet the adoption of such policies is politically difficult, because they break so radically with the dominant neoliberal understanding of the last thirty years of how the economy works, and how the relationships between private market actors and government agencies should be organised.

Another political obstacle is that political action has to be international as well as national. But international cooperation as Tony Payne explains is fragile, the main global institutions are weak, and the G20 powers have the capacity but not the political will to replace privatised corporate global governance with public global governance. The European Union, one of the key structures of the international market order, is under huge pressure from migration and from its dysfunctional single currency, as Nicola Phillips and Scott Lavery show in their chapters. The EU's failure to deliver serious economic and political reform and recover the project of social Europe to counterbalance the ordoliberal project of the single market has made it vulnerable to populist insurgencies. Failure to devise an effective response to migration and to the problems of growth and debt in the eurozone risks accelerating the disintegration of the Union, one of the major supports of the rules-based multilateral international market order.

So if a new crisis is coming what form might it take? There have been three distinct crisis phases since 2007–2008. The first was centred on finance with its epicentre in the heartlands of Anglo-America, the United States and the United Kingdom. That crisis was contained and managed but not resolved. The risks of another financial crash have not disappeared. The stock market has become once again hugely inflated, and even the modest restrictions that were placed on the banks are set to be dismantled by the Trump Administration. The potential of another collapse is possible, and this time the conventional tools like interest rates are no longer available to be deployed in response. The second crisis phase, a crisis of sovereign debt, centred on the eurozone in 2010–2012. That crisis was also contained and managed but not resolved, and as the renewed problems over Greece and Italy show may be about to erupt again. The third crisis phase 2013–2016 was centred on the emerging economies and rising powers, particularly, China and Brazil. Many of these economies which kept the international economy still growing strongly through the recessions in North America, Europe and Japan, have suffered sharp decelerations in growth since 2012 and rapidly rising levels of debt and signs of financial overheating. Again these problems have been contained and managed, but the pace of advance has notably slowed, and much of the earlier optimism about the unstoppable momentum of these economies has faded. All of them have become key parts of an interdependent international economy, and a major breakdown in one of them, particularly China, would affect the stability and prosperity of everyone. Matthew Bishop shows how the extraordinary success of the Chinese growth model is currently under strain, as the government seeks to manage a transition to an economy focused less on exports and more on domestic consumption, while maintaining the legitimacy and political monopoly of the Communist party. China needs to liquidate or restructure many of its large inefficient state-owned companies which have piled up huge debts, but may need to choose between a radical write-down of debt with all the attendant social and political risks both to China and the international economy, or a sharp slow-down in long-term growth. Japan in the same situation twenty five years earlier chose the latter.

These different crisis phases since 2008 reveal a world struggling to cope with excess debt, low growth, and regional and global inequalities. The international economy has increasingly been stuck in an impasse. Yet the national governments represented in the G20 have shown remarkable political resilience, both the authoritarian regimes and the western democracies. Many incumbent governments in the latter were toppled, but were always succeeded by another party from within the ruling policy consensus. In 2016, we entered a fourth phase of the crisis, a political crisis, which has the potential to reignite all the previous three crises and generate a truly transformative crisis. The political upheaval which is taking place is centred on a national-populism which rejects globalisation, neoliberalism and the rule of liberal cosmopolitan elites. One trigger has been the impact of austerity programmes in the last nine years, but another has been the desire to 'take back control' of borders, trade and production.

The epicentre of this fourth phase of the crisis has like the first been Anglo-America, with the vote for Brexit in June 2016 and the election of Donald Trump in November 2016. The two events have made breaches in the walls of the multilateral international market order but whether this will lead to lasting change and an unravelling of this order remains unclear. Trump promised many things in his campaign rallies—building a wall to keep Mexicans out, blocking entry of Muslims to the United States, imposing tariffs of up to 50 per cent on Chinese goods, renegotiating the North American Free Trade Agreement (NAFTA), not pursuing multilateral trade deals like the Transatlantic Trade and Investment Partnership (TTIP) or Trans-Pacific Partnership (TPP), compelling US multinationals to bring jobs back to the US, disengaging from alliances like NATO if other member states do not contribute more, supporting the fragmentation of the EU, and withdrawing from major international agreements, including on climate change and refugees.

Trump's proclamation of America First weaves together protectionist, nativist and isolationist themes, and as such is most closely in line with Europe's populists like Nigel Farage, Marine Le Pen and Geert Wilders. It is a new international movement, a global Tea Party. But despite the vote for Brexit in the UK the UK Government is still run by mainstream Conservatives who are not seeking to overturn the Western political and economic order. Although they are seeking to withdraw the UK from the EU, they claim to want a more 'global' and open Britain, despite the fact that many of those voting for Brexit were clearly voting for the opposite. In the UK, this means the revolution has still to happen, but in the US Trump's victory means that a representative of the national-populist right has broken through into government. There are still major obstacles to that happening in Europe, but it has become a real possibility, and the strength of the European order will continue to be tested at national elections over the coming years.

Trump has breached the wall but he has yet to show he has the strength or capacity to capture the citadel. The evidence is contradictory. His administration has become a battleground in which different factions struggle for the President's ear and to control the agenda.

This is normal. What is abnormal is what is being fought over. If Trump succeeds in 'draining the swamp', overcoming significant opposition in Congress, in the courts, in his own Cabinet, and in the deep state, he may be able to implement a radical agenda. On an optimistic scenario, he may break through the taboos of neoliberalism and globalisation and discover through trial and error a new national growth model for the US, and potentially for the world, while maintaining significant international cooperation. On a pessimistic scenario, his policies could signal the end of the post-war American order, and the unravelling of western prosperity with the onset of trade wars and a new world depression.

For the first time since 2008, the crisis has assumed a markedly political form, with a challenge to the political and economic assumptions of the international market order which has been dominant since 1945. But beyond the crises of debt, inequality and growth, there is another still greater crisis which overshadows it. In his chapter, Martin Craig sets out how capitalism is progressively eroding the ecological conditions which make its survival possible, while Peter Dauvergne argues that so far plans to take action, such as the 2015 Paris Agreement, while welcome, are not enough. The gains they secure will not keep pace with rising environmental costs. Although in the aftermath of the financial crash economic growth slowed in the more developed parts of the international economy, world economic growth still continued at 3 per cent per annum, and the cumulative impact of 200 years of industrialisation is having devastating impacts on climate, food, biodiversity and the pollution of air and water. In the search for ways forward to overcome the problems of debt, inequality and growth the solutions must also address this other truly existential crisis which looms ever more threateningly ahead. We do not have much time.

Index

C. Hay and T. Hunt (eds.), *The Coming Crisis*, Building a Sustainable Political Economy: SPERI Research & Policy,
DOI 10.1007/978-3-319-63814-0

CPI Antony Rowe
Eastbourne, UK
November 26, 2019